Process Capability Indices

Process Capability Indices

Samuel Kotz
University of Maryland, College Park, Maryland, USA

and

Norman L. Johnson
University of North Carolina, Chapel Hill, North Carolina, USA

CHAPMAN & HALL

London · Glasgow · New York · Tokyo · Melbourne · Madras

Published by Chapman & Hall, 2–6 Boundary Row, London SE1 8HN

Chapman & Hall, 2–6 Boundary Row, London SE1 8HN, UK

Blackie Academic & Professional, Wester Cleddens Road, Bishopbriggs, Glasgow G64 2NZ, UK

Chapman & Hall Inc., 29 West 35th Street, New York NY10001, USA

Chapman & Hall Japan, Thomson Publishing Japan, Hirakawacho Nemoto Building, 6F, 1–7–11 Hirakawa-cho, Chiyoda-ku, Tokyo 102, Japan

Chapman & Hall Australia, Thomas Nelson Australia, 102 Dodds Street, South Melbourne, Victoria 3205, Australia

Chapman & Hall India, R. Seshadri, 32 Second Main Road, CIT East, Madras 600 035, India

First edition 1993

© 1993 Chapman & Hall

Typeset in 11/13pt Times by Interprint Limited, Malta.
Printed in Great Britain by St Edmundsbury Press, Bury St Edmunds, Suffolk

ISBN   0 412 54380 X

A catalogue record for this book is available from the British Library

Library of Congress Cataloging-in-Publication data available.

∞ Printed on permanent acid-free text paper, manufactured in accordance with the proposed ANSI/NISO Z 39.48-199X and ANSI Z 39.48-1984

Contents

Foreword

Measures of process capability known as process capability indices (PCIs) have been popular for well over 20 years, since the capability ratio (CR) was popularized by Juran. Since that time we have seen the introduction of C_p, C_{pk}, C_{pm}, P_{pk} and a myriad of other measures.

The use of these measures has sparked a significant amount of controversy and has had a major economic impact (in many cases, negative) on industry. The issue does not generally lie in the validity of the mathematics of the indices, but in their application by those who believe the values are deterministic, rather than random variables. Once the variability is understood and the bias (if any) is known, the use of these PCIs can be more constructive.

As with statistics in general, it is imperative that the reader of this text has a working knowledge of statistical methods and distributional theory. This basic understanding is what has been lacking in the application of PCIs over the years.

It is hoped that this text will assist in moderating (as in this writer's case) some of the polarization that has occurred during the past decade over the use of PCIs and that we can work on process improvements in general, rather than focusing upon a single measure or index.

There are those of us who feel that the use of PCIs has been ineffective and should be discontinued. There are those who feel they have their use when used in conjunction with other measures and there are those who use PCIs as absolute measures. I would suggest that an understanding of the concepts in this text could bring the majority of viewpoints to a common ground.

Robert A. Dovich
December, 1992

Preface

The use and abuse of process capability indices (PCIs) have become topics of considerable controversy in the last few years. Although seemingly simple in formal use, experience has shown that PCIs lend themselves rather easily to ill-based interpretations. It is our hope that this monograph will provide background sufficient to allow for informal assessment of what PCIs can and cannot be expected to do.

We have tried to keep the exposition as elementary as possible, without obscuring its essential logical and mathematical components. Occasionally the more elaborate of the latter are moved to an appendix. We hope that Chapter 1 will provide a reminder of the concepts and techniques needed to study the book with profit.

We also hope, on the other hand, that more sophisticated researchers will find possible topics for further developmental work in the book.

Before venturing into this field of current controversy we had some misgivings about our abilities in this field. We have benefited substantially from the advice of practitioners (especially Mr. Robert A. Dovich, Quality Manager, Ingersoll Cutting Tool Company, Rockford, Illinois and Dr. K. Hung, Department of Finance, Western Washington University, Bellingham, Washington) but our approach to the subject is based primarily on our own experience of numerical applications in specific problems over many years and our prolonged study of distributional-theoretic aspects of statistics. We hope our efforts will contribute to lessening the gap between theory and practice.

We thank Dr. W.L. Pearn, of the National Chiao Tung University, Hsinchu, Taiwan, for his extensive collaboration in the early stages of this work, Mr. Robert A. Dovich for useful comments on our first version, and Ms. Nicki Dennis for her enthusiastic support of the project. Conversations and correspondence with Mr. Eric Benson and Drs. Russell A. Boyles, Richard Burke, Smiley W. Cheng, LeRoy A. Franklin, A.K. Gupta, Dan Grove, Norma F. Hubele, S. Kocherlakota, Peter R. Nelson, Leslie J. Porter, Barbara Price, Robert N. Rodriguez, Lynne Seymour and N.F. Zhang contributed to clarification of our ideas on the somewhat controversial subject matter of this monograph.

To paraphrase George Berkeley (1685–1753) our task and aim in writing this monograph was 'to provide hints to thinking men' and women who have determination 'and curiosity to go to the bottom of things and pursue them in their own minds'.

January 1993

Samuel Kotz   Norman L. Johnson
University of Maryland   University of North Carolina
College Park, MD   Chapel Hill, NC

1

Introduction

Process capability indices (PCIs) have proliferated in both use and variety during the last decade. Their widespread and often uncritical use may, almost inadvertently, have led to some improvements in quality, but also, almost certainly, have been the cause of many unjust decisions, which might have been avoided by better knowledge of their properties.

These seemingly innocuous procedures for determining 'process capability' by a single index were propagated mainly by over-zealous customers who viewed them as a panacea for problems of quality improvement. Insistence on rigid adherence to rules for calculating the indices C_p and C_{pk} (see Chapter 2) on a daily basis, with the goal of raising them above 1.333 as much as possible, caused a revolt among a number of influential and open-minded quality control statisticians. An extreme reaction against these indices took the form of accusations of 'statistical terrorism' – Dovich's letter (1991 a), Dovich (1991 b) and Burke *et al.* (1991) – and of unscrupulous manipulation and doctoring, and calls for their elimination – Kitska's letter (1991). There were also more moderate voices (Gunter, 1989) and more defenders (McCormick (in a letter, 1989 with reply by Gunter, 1989 b); Steenburgh, 1991; McCoy, 1991). However, the heated and intense debate on the advisability of continuation of the practical application of these indices – which took place during the first four months of 1991 – indicates that something is, indeed, wrong either with the indices themselves or the manner they had been 'sold' to quality control practitioners and the 'man on the shop floor'. Additional indices have been introduced in leading journals on quality control in the last few years (Chan *et al.*, 1988; Spiring, 1991; Boyles, 1991; Pearn *et al.*, 1992) in efforts to improve on C_p and C_{pk} by taking into account the possibility that the target (nominal) value may not correspond to the midpoint between the specification limits, and possible non-normality of the original process characteristic. The number of different capability indices has increased and so has confusion among practitioners, who have so far been denied an adequate and clear explanation of the meaning of the various indices, and more importantly the underlying assumptions. Their suspicion that a single number cannot fully capture the 'capability' of a process is well justified!

The purpose of this monograph is to clarify the statistical meaning of these indices by presenting the underlying theory, to investigate and compare, in some detail, the properties of various indices available in the literature, and finally to convince practitioners of the advisability of representing estimates of process capability by a pair of numbers rather by a single one, where the second (supplementary) number might provide an indicator of the sample variability (accuracy) of the primary index, which has commonly been, and sometimes still is, mistakenly accepted as a precise indicator of the process capability, without any reservations. This should defuse to some extent the tense atmosphere and the rigidity of opinions prevalent among quality controllers at present, and, perhaps, serve as an additional bridge of understanding between the consumer and the producer. It may even reduce the cost of production without compromising the quality. We consider this to be an important educational goal.

To benefit from studying this monograph, some basic knowledge of statistical methodology and in particular of the theory of a number of basic statistical distributions is essential. Although many books on quality control provide excellent descriptions of the normal distribution, the distributional (statistical) properties of the process capability indices involve several other important but much less familiar distributions. In fact, these indices give rise to new interesting statistical distributions which may be of importance and promise in other branches of engineering and sciences in general. For this reason most of this chapter will be devoted to a condensed, and as elementary as possible, study of those statistical distributions and their properties which are essential for a deeper understanding of the distributions of the process capability indices. Further analytical treatment may be found in books and articles cited in the bibliography. Specifically, we shall cover basic properties and characteristics of random variables and estimators in general, and then survey the properties of the following:

1. binomial and Poisson distributions
2. normal distributions
3. (central) chi-squared (and more generally gamma) distributions
4. non-central chi-squared distributions
5. beta distributions
6. t- and non-central t-distributions
7. folded normal distributions
8. mixture distributions
9. multinormal distributions

We also include, in section 1.12, a summary of techniques in statistical methodology which will be used in special applications.

1.1 FACTS NEEDED FROM DISTRIBUTION THEORY

We assume that the reader possesses, at least, a basic knowledge of the elements of probability theory. We commence with a condensed account of basic concepts to refresh the memory and establish notation. This is followed by a summary of results in distribution theory which will be used later in this monograph.

We interpret 'probability' as representing long-run relative frequency of an event rather than a 'degree of belief'. Probabilities are always in the interval $[0, 1]$. If an event always (never) occurs the probability is 1 (0). We will be dealing with probabilities of various types of events.

A random variable X represents a real-valued varying quantity such that the event '$X \leqslant x$' has a probability for each real number x. Symbolically

$$\Pr[X \leqslant x] = F_X(x). \tag{1.1}$$

The symbol $F_X(x)$ stands for the **cumulative distribution function** (CDF) of X (or more precisely its value at $X = x$). Note that it is a function (in the mathematical sense) of x, not of X. It represents a property of the random variable X, namely its distribution. For a *proper* random variable

$$\lim_{x \to -\infty} F_X(x) = 0 \quad \lim_{x \to \infty} F_X(x) = 1 \tag{1.2}$$

To represent quantities taking only integer values – for example, number of nonconforming items – we use discrete random variables. Generally these are variables taking only a number (finite, or countably infinite) of possible values $\{x_i\}$. In the cases we are dealing with the possible values are just non-negative integers and the distribution can be defined by the values

$$\Pr[X = j] \quad j = 0, 1, 2, \ldots$$

For such random variables the CDF is

$$F_X(x) = \sum_{j \leqslant x} \Pr[X = j] \tag{1.3}$$

To represent quantities such as length, weight, etc., which can (in principle) be thought of as taking any value over a range (finite or infinite interval) we use continuous random variables for which the derivative

$$f_X(x) = \frac{\mathrm{d}F_X(x)}{\mathrm{d}x} = F'_X(x) \tag{1.4}$$

of the CDF exists. The function $f_X(x)$ is the **probability density function** (PDF) of the random variable X. For a continuous random variable, X, for any real α and β, $(\beta > \alpha)$:

$$\begin{aligned}
\Pr[\alpha \leqslant X \leqslant \beta] &= \Pr[\alpha < X \leqslant \beta] = \Pr[\alpha \leqslant X < \beta] \\
&= \Pr[\alpha < X < \beta] \\
&= F_X(\beta) - F_X(\alpha) \\
&= \int_\alpha^\beta f_X(x)\,\mathrm{d}x
\end{aligned} \tag{1.5}$$

(For such a variable, $\Pr[X = \alpha] = \Pr[X = \beta] = 0$ for any α and β.) We will use x_ε to denote the (unique) solution of $F_x(x_\varepsilon) = \varepsilon$. The joint CDF of k random variables $X_1, \ldots, X_k$ is

$$\begin{aligned}
F_X(x) = F_{X_1,\ldots,X_k}(x_1, \ldots, x_k) &= \Pr[(X_1 \leqslant x_1) \cap \\
\cap (X_2 \leqslant x_2) \cap \cdots &\cap (X_k \leqslant x_k)] \\
&= \Pr\left[\bigcap_{j=1}^k (X_j \leqslant x_j) \right]
\end{aligned} \tag{1.6}$$

For continuous random variables, the joint PDF is

$$f_X(x) = f_{X_1,\ldots,X_k}(x_1, \ldots, x_k) = \frac{\partial^k F_X(x)}{\partial x_1 \partial x_2 \ldots \partial x_k} \tag{1.7}$$

(the kth mixed derivative with respect to each variable).

We denote the **conditional probability** of an event E_1 given an event E_2 by $\Pr[E_1|E_2]$. By definition, E_1 is independent of E_2 if $\Pr[E_1|E_2]=\Pr[E_1]$. Generally $\Pr[E_1$ and $E_2]$, denoted by $\Pr[E_1 \cap E_2]$, is equal to $\Pr[E_2]\Pr[E_1|E_2]=\Pr[E_1]\Pr[E_2|E_1]$.

Also $\Pr[E_1$ or E_2 or both], denoted by $\Pr[E_1 \cup E_2]$, is equal to $\Pr[E_1]+\Pr[E_2]-\Pr[E_1 \cap E_2]$.

If E_1 is independent of E_2 and $\Pr[E_2]>0$, then E_2 is independent of E_1 and we say that E_1 and E_2 are mutually independent.

Random variables X_1 and X_2 are (mutually) independent if the events $(X_1 \leqslant x_1)$ and $(X_2 \leqslant x_2)$ are mutually independent for all x_1 and x_2, so that

$$F_{X_1,X_2}(x_1,\ x_2)=F_{X_1}(x_1)F_{X_2}(x_2) \qquad (1.7)$$

The **expected value** of a random variable represents the long-run average of a sequence of observed values to which it corresponds. The expected value of a continuous random variable, X is

$$\xi=\mathrm{E}[X]=\int_{-\infty}^{\infty} x f_X(x)\,\mathrm{d}x \qquad (1.8\,a)$$

also denoted as $\mu_1'(X)$. Often the term 'mean value' or simply 'mean' is used.

More generally, if $g(X)$ is a function of X

$$\mathrm{E}[g(X)]=\int_{-\infty}^{\infty} g(x) f_X(x)\,\mathrm{d}x \qquad (1.9\,a)$$

For a discrete random variable

$$\xi=\mathrm{E}[X]=\sum_i x_i \Pr[X=x_i] \qquad (1.8\,b)$$

$$\mathrm{E}[g(X)]=\sum_i g(x_i)\Pr[X=x_i] \qquad (1.9\,b)$$

The **variance** of X is by definition

$$E[(X - E(X))^2] = E(X^2) - [E(X)]^2 \qquad (1.10\ a)$$

It is also written as

$$\text{var}(X) = \mu_2 = \sigma^2(X) = \sigma^2 \qquad (1.10\ b)$$

The square root of the variance is the **standard deviation**, usually denoted by $\sigma(X)$ or simply σ. This will be an important and frequently used concept in the sequel.

More generally, the rth **(crude) moment** of X is $E[X^r] = \mu_r'(X)$. Note that $\mu_1'(X) = E[X] = \xi$.

The rth **central moment** is by definition:

$$E[(X - E(X))^r] = \mu_r(X) \qquad (1.11)$$

The variance is the second central moment.

The **coefficient of variation** is the ratio of standard deviation to the expected value (C.V.$(X) = \sigma(X)/E[X]$), but it is often expressed as a percentage ($100\{\sigma(x)/E[X]\}\%$).

If X and Y are mutually independent, then $E[XY] = E[X]E[Y]$. For any two random variables, whether independent or not,

$$E[X + Y] = E[X] + E[Y]$$

The **correlation coefficient** between two random variables X and Y, is defined as

$$\rho_{XY} = \frac{E[(X - E[X])(Y - E[Y])]}{\sigma(X)\sigma(Y)}$$

The numerator, $E[(X - E[X])(Y - E[Y])]$ is called the **covariance** of X and Y, cov (X, Y).

Occasionally we need to use higher moments $(r > 2)$, in order to compute conventional measures of **skewness** (asymmetry) and **kurtosis** (flat- or peaked-'topness'). These are

$$(\beta_1(X))^{\frac{1}{2}} = \frac{\mu_3(X)}{\{\mu_2(X)\}^{\frac{3}{2}}} = \frac{\mu_3(X)}{\{\sigma(X)\}^3} \qquad (1.12\,a)$$

(the symbols $\alpha_3(X)$ or $\lambda_3(X)$ are also used for this parameter), and

$$\beta_2(X) = \frac{\mu_4(X)}{\{\mu_2(X)\}^2} = \frac{\mu_4(X)}{\{\sigma(X)\}^4} \qquad (1.12\,b)$$

(the symbols $\alpha_4(X)$ or $\{\lambda_4(X) + 3\}$ are also used for this parameter).

When there is no fear of confusion the '(X)' is often omitted, so we have for example

$$\sqrt{\beta_1} = \frac{\mu_3}{\sigma^3} \quad \beta_2 = \frac{\mu_4}{\sigma^4} \qquad (1.12\,c)$$

and

$$\text{C.V.}(X) = 100\sigma/\mu_1' \% \ (= 100\sigma/\xi\%) \qquad (1.13)$$

$\sqrt{\beta_1}$ and β_2 are referred to as **moment ratios** or **shape factors** (because they reflect the shape of the distribution rather than its location – measured by ξ – or its scale – measured by σ). Neither $\sqrt{\beta_1}$ nor β_2 is changed if X is replaced by $Y = aX + b$, when a and b are constants and $a > 0$. If $a < 0$, the sign of $\sqrt{\beta_1}$ is reversed, but β_2 remains unchanged. If $\sqrt{\beta_1} < 0$ the distribution is said to be **negatively skew**, if $\sqrt{\beta_1} > 0$ the distribution is said to be **positively skew**. For any normal distribution, $\sqrt{\beta_1} = 0$ and $\beta_2 = 3$. If $\beta_2 < 3$ the distribution is said to be **platykurtic** ('flat-topped'); if $\beta_2 > 3$ the distribution is said to be **leptokurtic** ('peak-topped').

β_2 cannot be less than 1. It is not possible to have $\beta_2 - \beta_1 - 1 < 0$. (If $\beta_2 - \beta_1 - 1 = 0$ the distribution of X reduces to the 'two-point' distribution with $\Pr[X = a] = 1 - \Pr[X = b]$ for arbitrary $a \neq b$.)

We state here without proof, results which we will use in later chapters. If $X_1, X_2, \ldots, X_n$ are mutually independent with common expected value, ξ, variance, σ^2, and kurtosis shape factor, β_2 then for the arithmetic mean,

$$\bar{X} = n^{-1} \sum_{i=1}^{n} X_i$$

$$E[\bar{X}] = \xi \quad \mathrm{var}(\bar{X}) = \frac{\sigma^2}{n} \qquad (1.14\,a)$$

For the sample variance

$$S^2 = \frac{1}{n-1} \sum_{i=1}^{n} (X_i - \bar{X})^2$$

we have

$$E[S^2] = \sigma^2 \qquad (1.14\,b)$$

$$\mathrm{var}(S^2) = \left(\beta_2 - \frac{n-3}{n-1} \right) \sigma^4 \qquad (1.14\,c)$$

We also note the simple but useful inequality

$$E[|X|] \geqslant |E[X]| \qquad (1.14\,d)$$

(Since $E[|X|] \geqslant E[X]$ and $E[|X|] = E[|-X|] \geqslant E[-X] = -E[X]$.) Note also the identity

$$\min(a, b) = \tfrac{1}{2}(|a+b| - |a-b|) \qquad (1.14\,e)$$

1.2 APPROXIMATIONS WITH STATISTICAL DIFFERENTIALS ('DELTA METHOD')

Much of our analysis of properties of PCIs will be based on the assumption that the process distribution is normal. In this case we can obtain exact formulas for moments of sample estimators of the PCIs we will discuss. In Chapter 4, however, we address problems arising from non-normality, and in most cases we will not have exact formulas for distributions of estimators of PCIs, or even for the moments. It is therefore useful to have some way of approximating these values, one of which we will now describe. (Even in the case of normal process distributions, our approximations may be useful when they replace a cumbersome exact formula by a simpler, though only approximate formula. It is, of course, desirable to reassure ourselves, so far as possible, with regard to the accuracy of the latter.)

Suppose we have to deal with the distribution of the ratio of two random variables

$$G = \frac{A}{B}$$

and

$$E[A] = \alpha \quad E[B] = \beta$$
$$\mathrm{var}(A) = \sigma_A^2 \quad \mathrm{var}(B) = \sigma_B^2$$
$$\mathrm{cov}(A, B) = \rho_{AB} \sigma_A \sigma_B$$

i.e. the correlation coefficient between A and B is ρ_{AB}.

To approximate the expected value of G^r we write

$$G^r = \left(\frac{A}{B}\right)^r = \left\{\frac{\alpha + (A - \alpha)}{\beta + (B - \beta)}\right\}^r = \left(\frac{\alpha}{\beta}\right)^r \left(1 + \frac{A - \alpha}{\alpha}\right)^r \left(1 + \frac{B - \beta}{\beta}\right)^{-r}$$

$$(1.15)$$

The deviations $A-\alpha$, $B-\alpha$ of A and B from their expected values are called **statistical differentials**. Clearly

$$E[A-\alpha]=0 \quad E[(A-\alpha)^2]=\sigma_A^2 \quad E[(A-\alpha)(B-\beta)]=\rho_{AB}\sigma_A\sigma_B$$

(The notation $A-\alpha=\delta(A)$; $B-\beta=\delta(B)$ is sometimes used, giving rise to the alternative name: **delta method**.)

We now expand formally the second and third components of the right-hand side of (1.15). If r is an integer, the expansion of the second component terminates, but that of the third component does not. Thus we have (with $r=1$)

$$G=\frac{\alpha}{\beta}\left(1+\frac{A-\alpha}{\alpha}\right)\left\{1-\frac{B-\beta}{\beta}+\left(\frac{B-\beta}{\beta}\right)^2-\cdots\right\} \quad (1.16)$$

The crucial step in the approximation is to suppose that we can terminate the last expansion and neglect the remaining terms. This is unlikely to lead to satisfactory results unless $|(B-\beta)/\beta|$ is small. Indeed, if $|(B-\beta)/\beta|$ exceeds 1, the series does not even converge.

Setting aside our misgivings for the present, we take expected values of both sides of (1.16), obtaining

$$E[G]\cong\frac{\alpha}{\beta}\left(1-\frac{\rho_{AB}\sigma_A\sigma_B}{\alpha\beta}+\frac{\sigma_B^2}{\beta^2}\right) \quad (1.17\,a)$$

A similar analysis leads to

$$\text{var}(G)\cong\left(\frac{\alpha}{\beta}\right)^2\left(\frac{\sigma_A^2}{\alpha^2}-\frac{2\rho_{AB}\sigma_A\sigma_B}{\alpha\beta}+\frac{\sigma_B^2}{\beta^2}\right) \quad (1.17\,b)$$

Formulas (1.17a) and (1.17b) can be written conveniently in terms of the coefficients of variation of A and B (see p. 7).

$$E[G]\cong\frac{\alpha}{\beta}[1-\rho_{AB}\{C.V.(A)C.V.(B)\}+\{C.V.(B)\}^2] \quad (1.17\,c)$$

$$\text{var}(G) \cong \left(\frac{\alpha}{\beta}\right)^2 [\{C.V.(A)\}^2 - 2\rho_{AB}\{C.V.(A)C.V.(B)\}$$

$$+ \{C.V.(B)\}^2] \qquad (1.17\,d)$$

Another application of the method is to approximate the expected value of a function $g(A)$ in terms of moments of A. Expanding in a Taylor series

$$g(A) = g(\alpha + A - \alpha) = g(\alpha) + (A - \alpha)g'(\alpha) + \tfrac{1}{2}(A - \alpha)^2 g''(\alpha) + \cdots$$
$$\cong g(\alpha) + (A - \alpha)g'(\alpha) + \tfrac{1}{2}(A - \alpha)^2 g''(\alpha)$$

Taking expected values of both sides

$$E[g(A)] \cong g(\alpha) + \tfrac{1}{2}g''(\alpha)\sigma_A^2 \qquad (1.18\,a)$$

Similar analysis leads to

$$\text{var}(g(A)) \cong \{g'(\alpha)\}^2 \text{var}(A) = \{g'(\alpha)\}^2 \sigma_A^2 \qquad (1.18\,b)$$

Example (see Sections 1.4 and 1.5 for notation)

Suppose we want to approximate the expected value and variance of

$$\left\{\sum_{i=1}^{n} (X_i - \bar{X})^2\right\}^{-\frac{1}{2}}$$

where $X_1, \ldots, X_n$ are independent $N(\xi, \sigma^2)$ random variables and

$$\bar{X} = \frac{1}{n} \sum_{i=1}^{n} X_i.$$

We take

$$\sum_{i=1}^{n} (X_i - \bar{X})^2$$

as our 'A', and $g(A) = A^{-\frac{1}{2}}$.

From (1.31) we will find that

$$\sum_{i=1}^{n} (X_i - \bar{X})^2 \quad \text{is distributed as} \quad \chi_{n-1}^2 \sigma^2$$

Hence

$$\alpha = \mathrm{E}[A] = (n-1)\sigma^2$$

and

$$\sigma_A^2 = 2(n-1)\sigma^4$$

Also

$$g'(A) = -\tfrac{1}{2}A^{-\frac{3}{2}} \quad g''(A) = \tfrac{3}{4}A^{-\frac{5}{2}}$$

Hence

$$\mathrm{E}[A^{-\frac{1}{2}}] \cong \{(n-1)\sigma^2\}^{-\frac{1}{2}} + \tfrac{1}{2}\{2(n-1)\sigma^4\}\{\tfrac{3}{4}(n-1)^{-\frac{5}{2}}\sigma^{-5}\}$$

$$\cong (n-1)^{-\frac{1}{2}}\left\{1 + \frac{3}{4(n-1)}\right\}\sigma^{-1} \qquad (1.19\,a)$$

and

$$\mathrm{var}(A^{-\frac{1}{2}}) \cong 2(n-1)\sigma^4 \times \left(\frac{1}{4\alpha^3}\right)$$

$$= 2(n-1)\sigma^4 \times \frac{1}{4(n-1)^3}\sigma^{-6}$$

$$= \frac{1}{2(n-1)^2 \sigma^2} \qquad (1.19\,b)$$

Remembering the formulas $(1.14\,b, c)$ for $E[S^2]$ and $\text{var}(S^2)$ we find that

$$\sigma(S^{-1}) \cong \frac{1}{2}\left(\beta_2 - \frac{n-3}{n-1}\right)^{\frac{1}{2}} \sigma^{-1} \qquad (1.19\,c)$$

for any distribution.

1.3 BINOMIAL AND POISSON DISTRIBUTIONS

Suppose that in a sequence of n observatiòns we observe whether an event E (e.g. $X \leqslant x$) occurs or not. We can represent the number of occasions on which E occurs by a random variable X. To establish an appropriate distribution for X, we need to make certain assumptions. The most common set of assumptions is:

1. for each observation the probability of occurrence of E has the same value, p, say – symbolically – $\Pr[E] = p$; and
2. the set of results of n observations are mutually independent.

Under the above stated conditions

$$\Pr[X = k] = \binom{n}{k} p^k (1-p)^{n-k} \quad \text{for } k = 0, 1, \ldots, n \qquad (1.20)$$

where

$$\binom{n}{k} = \frac{n!}{k!(n-k)!} = \frac{n^{(k)}}{k!}$$

(Recall that $n! = 1 \times 2 \times \cdots \times n$ and $n^{(k)} = n \times (n-1) \times \cdots \times (n-k+1)$.)

Denoting $q = 1 - p$ (a common notation in the literature), we observe that the quantities

$$\Pr[X=0], \quad \Pr[X=1], \quad \Pr[X=2], \ldots, \quad \Pr[X=n]$$

are the successive terms in the binomial expansion of $(q+p)^n$.

The distribution defined by (1.20) is called a **binomial distribution with parameters n and p**, briefly denoted by Bin(n, p).

The expected value and variance of the distribution are

$$\mu'_1(X) = E[X] = np$$
$$\mu_2(X) = \text{Var}(X) = npq$$

respectively.

If n is increased indefinitely and p decreased, keeping the expected value np constant $(= \theta$ say) then

$$\Pr[X=k] \quad \text{approaches to} \quad \theta^k \frac{\exp(-\theta)}{k!}$$

The distribution defined by

$$\Pr[X=k] = \theta^k \frac{\exp(-\theta)}{k!} \quad k=0, 1, \ldots$$

is a **Poisson distribution** with expected value θ, denoted symbolically as:

$$X \sim \text{Poi}(\theta) \tag{1.21}$$

For this distribution, $E[X] = \theta$ and $\text{var}(X) = \theta$.

The *multinormal distribution* extends the binomial distribution to represent that of numbers of observations $N_1, \ldots, N_k$ of mutually exclusive events $E_1, \ldots, E_k$ in n independent trials with $\Pr[E_j] = p_j$ $(j = 1, \ldots, k)$ at each trial. We have

$$\Pr[(N_1 = n_1) \cap \cdots \cap (N_k = n_k)] = n! \prod_{j=1}^{k} \left(\frac{p_j^{n_j}}{n_j} \right) \qquad (1.22)$$

with

$$\sum_{j=1}^{k} n_j = n, \qquad \sum_{j=1}^{k} p_j = 1$$

1.4 NORMAL DISTRIBUTIONS

A random variable U has a standard (or unit) *normal* distribution if its CDF is

$$F_U(u) = \frac{1}{\sqrt{2\pi}} \int_{-\infty}^{u} \exp(-\tfrac{1}{2}t^2) \, dt = \Phi(u) \qquad (1.23\,a)$$

and its PDF is

$$f_U(u) = \frac{1}{\sqrt{2\pi}} \exp(-\tfrac{1}{2}u^2) = \varphi(u) \qquad (1.23\,b)$$

(We shall use the notation $\Phi(u), \varphi(u)$ throughout this book.)

This distribution is also called a **Gaussian distribution**, and occasionally a **bell-shaped distribution**. The expected value and the variance of U are 0 and 1 respectively.

The CDF and PDF of $X = \xi + \sigma U$ for $\sigma > 0$ and $\xi \neq 0$ are

$$F_X(x) = \frac{1}{\sigma\sqrt{2\pi}} \int_{-\infty}^{x} \exp\left[-\frac{1}{2}\left(\frac{t-\xi}{\sigma}\right)^2\right] dt = \Phi\left(\frac{x-\xi}{\sigma}\right)$$

$$(1.24\,a)$$

and

$$f_X(x) = \frac{1}{\sigma\sqrt{2\pi}} \exp\left[-\frac{1}{2}\left(\frac{x-\xi}{\sigma}\right)^2\right] = \frac{1}{\sigma}\varphi\left(\frac{x-\xi}{\sigma}\right) \quad (1.24\,b)$$

respectively. We write symbolically $X \sim N(\xi, \sigma^2)$.

The expected value of X is $E[X] = \xi$ and the variance of X is σ^2. The standard deviation is σ.

The distribution of U is symmetric about zero, i.e. $f_U(-u) = f_U(u)$. The distribution of X is symmetric about its expected (mean) value ξ, i.e.

$$f_X(\xi + x) = f_X(\xi - x) \tag{1.25}$$

There are extensive tables of $\Phi(u)$ available which are widely used. Typical values are shown in Table 1.1.

Table 1.1   Values $\Phi(u)$, the standardized normal CDF

u	$\Phi(u)$	u	$\Phi(u)$
0	0.5000	0.6745	0.7500
0.5	0.6915	1.2816	0.9000
1.0	0.8413	1.6449	0.9500
1.5	0.9332	1.9600	0.9750
2.0	0.97725	2.3263	0.9900
2.5	0.9938	2.5758	0.9950
3.0	0.99865	3.0902	0.9990
		3.2905	0.9995

Note that $u_1 = 1.645$, and $u_2 = 1.96$ correspond to $\Phi(u_1) = 0.95$ and $\Phi(u_2) = 0.975$ respectively, i.e. the area under the normal curve from $-\infty$ up to 1.645 is 95% and that from $-\infty$ up to 1.96 is 97.5%. Hence by symmetry the area between -1.645 and 1.645 is 90% and between -1.96 and 1.96, it is 95%. For applications in process capability indices we note the area under the normal curve between -3 and 3 to be 99.73%. Thus the area outside these limits is 0.0027 (an extremely small proportion).

If $X_1, X_2, \ldots, X_n$ are independent random variables with common $N(\xi, \sigma^2)$ distribution, then

$$\bar{X} = \frac{1}{n}(X_1 + X_2 + \cdots + X_n) = \frac{\sum_{i=1}^{n} X_i}{n} \sim N\left(\xi, \frac{\sigma^2}{n}\right) \quad (1.26)$$

1.5 (CENTRAL) CHI-SQUARED DISTRIBUTIONS AND NON-CENTRAL CHI-SQUARED DISTRIBUTIONS

1.5.1 Central chi-squared distributions

If $U_1, U_2, \ldots, U_v$ are mutually independent unit normal variables, the distribution of

$$\chi_v^2 = U_1^2 + U_2^2 + \cdots + U_v^2$$

has the PDF

$$f_{\chi_v^2}(x) = \left[2^{\frac{1}{2}v}\Gamma\left(\frac{v}{2}\right)\right]^{-1} x^{\frac{1}{2}v-1} \exp\left(-\frac{x}{2}\right) \quad x > 0 \quad (1.27)$$

where $\Gamma(\alpha)$ is the **Gamma function** defined by

$$\Gamma(\alpha) = \int_0^\infty y^{\alpha-1} \exp(-y)\, dy \quad (1.28)$$

(Gamma function is a generalization of factorials; since $\Gamma(n+1)=n!$ for any positive integer n.)

This is a χ^2 (Chi-squared) distribution with v degrees of freedom. Symbolically we write

$$\sum_{i=1}^{v} U_i^2 \sim \chi_v^2$$

The expected value of a χ_v^2 variable is v and its variance is $2v$.

If $V_1, V_2, \ldots, V_v$ are mutually independent normal variables each with expected value zero and standard deviation σ (variance σ^2); then

$$W = V_1^2 + V_2^2 + \cdots + V_v^2$$

has the PDF

$$f_W(w) = \left[2^{\frac{1}{2}v}\sigma^v \Gamma\left(\frac{v}{2}\right) \right] w^{\frac{1}{2}v-1} \exp\left(\frac{-w}{2\sigma^2}\right) \quad w>0 \qquad (1.29)$$

Symbolically

$$W \sim \sigma^2 \chi_v^2$$

W has expected value $\sigma^2 v$ and variance $2\sigma^4 v$. Generally

$$\mu_r'(\chi_v^2) = E[(\chi_v^2)^r] = \frac{2^r \Gamma(\frac{1}{2}v+r)}{\Gamma(\frac{1}{2}v)}$$

$$(= v(v+2)\cdots(v+2\overline{r-1}) \text{ if } r \text{ is a positive integer}).$$

If $r \leqslant \frac{1}{2}v$, then $\mu_r'(\chi_v^2)$ is infinite.

Generally a distribution with PDF

$$f_X(x) = \frac{1}{\beta^\alpha \Gamma(\alpha)} x^{\alpha-1} \exp\left(\frac{-x}{\beta}\right) \quad x>0; \alpha>0, \beta>0 \qquad (1.30)$$

is called a **gamma (α, β) distribution**.

A chi-square random variable with v degrees of freedom (χ_v^2) has a gamma $(\frac{1}{2}v, 2)$ distribution.

If $X_1, X_2, \ldots, X_n$ are independent $N(\xi, \sigma^2)$ random variables, then 'the sum of squared deviations from the mean' is distributed as $\chi_{n-1}^2 \sigma^2$. Symbolically (c.f. p. 13)

$$\sum_{i=1}^{n} (X_i - \bar{X})^2 \sim \chi_{n-1}^2 \sigma^2 \tag{1.31}$$

Also

$$\sum_{i=1}^{n} (X_i - \bar{X})^2 \quad \text{and} \quad \bar{X} = n^{-1} \sum_{i=1}^{n} X_i$$

(which is distributed $N(\xi, \sigma^2/n)$) are mutually independent.

1.5.2 Non-central chi-squared distributions

Following a similar line of development to that in Section 1.5.1, the distribution of

$$\chi_v'^2(\delta^2) = (U_1 + \xi_1)^2 + (U_2 + \xi_2)^2 + \cdots + (U_v + \xi_v)^2$$

where $U_i \sim N(0, 1)$ with

$$\sum_{i=1}^{v} \xi_i^2 = \delta^2$$

has the PDF

$$f_{\chi_v'^2(\delta^2)}(x) = \sum_{j=0}^{\infty} \left[\left(\frac{\delta^2}{2} \right)^j \frac{1}{j!} \right] \exp\left(\frac{-\delta^2}{2} \right)$$

$$\times \left\{ 2^{\frac{1}{2}v+j}\Gamma\left(j+\frac{v}{2}\right) \right\}^{-1} x^{\frac{1}{2}v+j-1} \exp\left(\frac{-x}{2}\right)$$

$$= \sum_{j=0}^{\infty} P_j\left(\frac{\delta^2}{2}\right) f_{\chi^2_{v+2j}}(x) \tag{1.32}$$

where

$$P_j\left(\frac{\delta^2}{2}\right) = \left(\frac{\delta^2}{2}\right)^j \frac{\exp\left(\dfrac{-\delta^2}{2}\right)}{j!} = \Pr[Y=j]$$

if Y has a Poisson distribution with expected value $\frac{1}{2}\delta^2$.

This is a non-central chi-squared distribution with v degrees of freedom and non-centrality parameter δ^2. Note that the distribution depends only on the sum δ^2 and not on the individual values of the summands ξ_i^2.

The distribution is a mixture (see section 1.9) of χ^2_{v+2j} distributions with weights $P_j(\frac{1}{2}\delta^2)$ $(j=0,1,2,\ldots)$.

Symbolically we write

$$\chi'^2_v(\delta^2) \sim \chi^2_{v+2J} \bigwedge_J \text{Poi}\left(\frac{\delta^2}{2}\right) \tag{1.33}$$

The expected value and variance of the $\chi'^2_v(\delta^2)$ distribution are $(v+\delta^2)$ and $2(v+2\delta^2)$ respectively.

Note that

$$\sum_{h=1}^{k} \chi'^2_{v_h}(\delta_h^2) \sim \chi'^2_v\left(\sum_{h=1}^{k} \delta_h^2\right) \quad \text{with } v = \sum_{h=1}^{k} v_h \tag{1.34}$$

if the χ'^2s in the summation are mutually independent.

1.6 BETA DISTRIBUTIONS

A random variable Y has a **standard beta (α, β) distribution** if its PDF is

$$f_Y(y) = \frac{1}{B(\alpha, \beta)} \, y^{\alpha-1}(1-y)^{\beta-1} \quad 0 \leqslant y \leqslant 1 \text{ and } \alpha, \beta > 0$$

$$(1.35\,a)$$

where $B(a, b)$ is the **beta function**, $= \Gamma(\alpha)\Gamma(\beta)/\Gamma(\alpha+\beta)$. Thus

$$f_Y(y) = \frac{\Gamma(\alpha+\beta)}{\Gamma(\alpha)\Gamma(\beta)} \, y^{\alpha-1}(1-y)^{\beta-1} \quad 0 \leqslant y \leqslant 1 \qquad (1.35\,b)$$

If $\alpha = \beta = 1$, the PDF is constant and this special case is called a standard **uniform distribution**.

If X_1 and X_2 are mutually independent and $X_j \sim \chi^2_{v_j}$ $(j = 1, 2)$ then

$$X_1 + X_2 \sim \chi^2_{v_1 + v_2} \quad \text{and} \quad \frac{X_1}{X_1 + X_2} \sim \text{Beta}\left(\frac{v_1}{2}, \frac{v_2}{2}\right) \qquad (1.36)$$

moreover $X_1/(X_1 + X_2)$ and $X_1 + X_2$ are mutually independent. (These results will be of importance when we shall derive distributions of various PCIs.)

The rth (crude) moment of Y is

$$\mu'_r(Y) = E[Y^r] = \frac{\Gamma(\alpha+r)\Gamma(\alpha+\beta)}{\Gamma(\alpha)\Gamma(\alpha+\beta+r)} = \frac{\alpha^{[r]}}{(\alpha+\beta)^{[r]}} \qquad (1.37)$$

where $a^{[b]} = a(a+1)\cdots(a+b-1)$ is the bth ascending factorial of a. If $r \leqslant -\alpha$, $\mu'_r(Y)$ is infinite.

In particular,

$$E[Y]=\mu'_1(Y)=\frac{\alpha}{\alpha+\beta}$$ (1.38 *a*)

$$E[Y^2]=\mu'_2(Y)=\frac{\alpha(\alpha+1)}{(\alpha+\beta)(\alpha+\beta+1)}$$ (1.38 *b*)

and hence

$$\text{var}(Y)=E(Y^2)-\{E(Y)\}^2=\frac{\alpha\beta}{(\alpha+\beta)^2(\alpha+\beta+1)}$$ (1.38 *c*)

We note here the following definitions:

1. the (**complete**) **beta function**;

$$B(\alpha,\beta)=\int_0^1 y^{\alpha-1}(1-y)^{\beta-1}\,dy \quad \alpha,\beta>0$$

2. the **incomplete beta function**

$$B_b(\alpha,\beta)=\int_0^b y^{\alpha-1}(1-y)^{\beta-1}\,dy \quad 0<b<1$$

3. the **incomplete beta function ratio**

$$I_b(\alpha,\beta)=\frac{B_b(\alpha,\beta)}{B(\alpha,\beta)}$$

If Y has the PDF (1.35 *a*) then the CDF is

$$F_Y(y)=I_y(\alpha,\beta) \quad 0<y<1$$

1.7 t-DISTRIBUTIONS AND NON-CENTRAL t-DISTRIBUTIONS

If $V \sim N(0, \sigma^2)$ and $S \sim \chi_v \sigma / \sqrt{v}$ are mutually independent then $T = V/S = U\sigma/S$ has the PDF

$$f_Y(y) = \frac{\Gamma(\frac{1}{2}(v+1))}{(v\pi)^{\frac{1}{2}}\Gamma(\frac{1}{2}v)} \left(1 + \frac{t^2}{v}\right)^{-\frac{1}{2}(v+1)} \qquad (1.39)$$

This is known as a **t-distribution with v degrees of freedom.** The t-distribution is symmetrical about zero; thus $E[T] = 0$. Note that $V = U\sigma$, where U is a unit normal variable. Hence

$$E[T^r] = E[v^{\frac{1}{2}r}U^r\chi_v^{-r}] = v^{\frac{1}{2}r}E[U^r]E[(\chi_v^2)^{-\frac{1}{2}r}] \qquad (1.40)$$

(which is also the rth central moment in this case since $E[T] = 0$). The variance of T is thus

$$\mathrm{var}(T) = E[T^2] = v \cdot 1 \cdot (v-2)^{-1} = \frac{v}{v-2} \qquad (1.41)$$

A common and widely used statistic giving rise to a t-distribution is

$$\frac{\sqrt{n}(\bar{X} - \xi)}{\left\{\dfrac{1}{n-1} \displaystyle\sum_{i=1}^{n} (X_i - \bar{X})^2\right\}^{\frac{1}{2}}} \qquad (1.42)$$

where $\bar{X}$ is 'sample mean' as defined in section 1.2 and $X_1, \ldots, X_n$ are mutually independent $N(\xi, \sigma^2)$ random variables. This statistic has a t-distribution with $(n-1)$ degrees of freedom.

If T is defined as above, but expected value of V is not 0 but δ, so that $V = (U + \delta/\sigma)\sigma$ with $U \sim N(0, 1)$ then the distribution of

$$T = \frac{U + \dfrac{\delta}{\sigma}}{\dfrac{\chi_v}{\sqrt{v}}} \qquad (1.43)$$

is called a **non-central t-distribution** with v degrees of freedom and non-centrality parameter δ/σ.

The expected value of T^r (the rth moment of T about the origin or the rth crude moment) is

$$E[T^r] = v^{\frac{1}{2}r} E\left[\left(U + \frac{\delta}{\sigma}\right)^r\right] \times E[(\chi_v^2)^{-\frac{1}{2}r}] \qquad (1.44)$$

In particular

$$E[T] = \left(\frac{v}{2}\right)^{\frac{1}{2}} \frac{\Gamma\left(\dfrac{v-1}{2}\right)}{\Gamma\left(\dfrac{v}{2}\right)} \frac{\delta}{\sigma} \qquad (1.45\ a)$$

and

$$E[T^2] = v\left(1 + \frac{\delta^2}{\sigma^2}\right)\frac{1}{v-2} \qquad (1.45\ b)$$

If the χ_v in the denominator of T is replaced by a noncentral $\chi_v'(\lambda)$ the resulting variable

$$T' = \frac{U + \dfrac{\delta}{\sigma}}{\dfrac{\chi_v'(\lambda)}{\sqrt{v}}}$$

has a double non-central t distribution with parameters δ/σ and λ, and ν degrees of freedom.

1.8 FOLDED DISTRIBUTIONS

Generally the distribution of $\{\theta+|X-\theta|\}$ is called the distribution of X folded (upwards) about θ; 'upwards' is usually omitted. It is used if it is necessary to distinguish it from the folded (downwards) about θ distribution which is the distribution of $\{\theta-|X-\theta|\}$.

The **folded normal distribution** formed by folding $N(\xi, \sigma^2)$ upwards about θ has the PDF

$$\frac{1}{\sigma\sqrt{2\pi}}\left\{\exp\left[-\frac{1}{2}\left(\frac{x-\xi}{\sigma}\right)^2\right]+\exp\left[-\frac{1}{2}\left(\frac{2\theta-x-\xi}{\sigma}\right)^2\right]\right\} \quad x>\theta$$

$$(1.46\,a)$$

If $\theta=0$, the variable is $|X|=\sigma|\nu+\delta|$ with expected value

$$\mu_1'=\left(\frac{2}{\pi}\right)^{\frac{1}{2}}\exp\left(-\frac{\delta^2}{2}\right)-\xi\{1-2\Phi(-\delta)\} \qquad (1.46\,b)$$

where $\delta=\xi/\sigma$ and the variance is

$$\mu_2=\xi^2+\sigma^2-\mu_1'^2 \qquad (1.46\,c)$$

Also

$$\mu_3=2\left[\mu_1'^3-\xi^2\mu_1'-\frac{\sigma^2}{\sqrt{2\pi}}\exp\left(-\frac{\delta^2}{2}\right)\right] \qquad (1.46\,d)$$

and

$$\mu_4 = \xi^4 + 6\xi^2\sigma^2 + 3\sigma^4 + \frac{8\sigma^3}{\sqrt{2\pi}}\mu_1' \exp\left(-\frac{\delta^2}{2}\right)$$
$$+ 2(\xi^2 - 3\sigma^2)\mu_1'^2 - 3\mu_1'^4 \qquad (1.46\,e)$$

(Elandt (1961)).

The folded (upwards) beta distribution formed by folding beta (α, β) about $\frac{1}{2}$ has the PDF

$$\frac{1}{B(\alpha + \beta)}\{x^{\alpha-1}(1-x)^{\beta-1} + x^{\beta-1}(1-x)^{\alpha-1}\} \quad \tfrac{1}{2} < x < 1$$
$$(1.47)$$

The corresponding folded (downwards) beta distribution has a PDF with the same mathematical form but valid over the interval $0 \leqslant x \leqslant \frac{1}{2}$ rather than $\frac{1}{2} \leqslant x \leqslant 1$.

1.9 MIXTURE DISTRIBUTIONS

If $F_1(x), F_2(x), \ldots, F_k(x)$ are CDFs then

$$F_X(x) = \sum_{j-1}^{k} \omega_j F_j(x) \quad \omega_j > 0, \ \sum_{j=1}^{k} \omega_j = 1 \qquad (1.48)$$

is the CDF of a mixture distribution, with k components $F_1(x), \ldots, F_k(x)$ and weights $\omega_1, \ldots, \omega_k$ respectively.

If the PDFs $f_j(x) = F_j'(x)$ exist, then the mixture distribution has PDF

$$f_X(x) = \sum_{j=1}^{k} \omega_j f_j(x) \qquad (1.49)$$

The rth crude moment of X is

$$\mu'_r(X) = \sum_{j=1}^{k} \omega_j \mu'_{jr} \tag{1.50}$$

where μ'_{jr} is the rth crude moment of the jth component distribution.

In particular

$$E[X] = \sum_{j=1}^{k} \omega_j \xi_j \tag{1.51 a}$$

and

$$E[X^2] = \sum_{j=1}^{k} \omega_j (\xi_j^2 + \sigma_j^2) \tag{1.51 b}$$

so that

$$\mathrm{var}(X) = \sum_{j=1}^{k} \omega_j(\xi_j^2 + \sigma_j^2) - \left(\sum_{j=1}^{k} \omega_j \xi_j \right)^2$$

$$= \sum_{j=1}^{k} \omega_j \sigma_j^2 + \sum_{j=1}^{k} \omega_j (\xi_j - \bar{\xi})^2 \tag{1.51 c}$$

where ξ_j, σ_j are the expected value and standard deviation, respectively, of the jth component distribution and

$$\bar{\xi} = \sum_{j=1}^{k} \omega_j \xi_j \ (= E[X])$$

If $F_j(x) = G(x, \theta_j)$, where $G(x, \theta_j)$ is a mathematical function of x and θ_j (e.g. a function corresponding to the $N(0, \sigma_j^2)$ distribution), we may regard a parameter Θ as having the discrete distribution

$$\Pr[\Theta = \theta_j] = \omega_j \quad j = 1, \ldots, k$$

More generally, Θ might have a continuous distribution with PDF $f_\Theta(t)$. Then the CDF of the mixture distribution is

$$F(x) = \int_{-\infty}^{\infty} f_\Theta(t) G(x, t) \, dt \qquad (1.52\,a)$$

If

$$\frac{dG(x, t)}{dx} = g(x, t)$$

is the PDF corresponding to $G(x, t)$, the PDF of the mixture distribution is

$$f(x) = \int_{-\infty}^{\infty} g(x, t) f_\Theta(t) \, dt \qquad (1.52\,b)$$

We denote 'mixture' symbolically by $\bigwedge_\Theta$ where Θ is called the mixing parameter. For example the distribution constructed by mixing $N(\xi, \sigma^2)$ distributions with σ^{-2} having a gamma (α, β) distribution will be represented symbolically as

$$N(\xi, \sigma^2) \bigwedge_{\sigma^{-2}} \text{gamma } (\alpha, \beta) \qquad (1.53)$$

1.10 MULTINORMAL (MULTIVARIATE NORMAL) DISTRIBUTIONS

The p random variables $\mathbf{X} = (x_1, \ldots, x_p)'$ have a joint **multivariate normal**, or **multinormal distribution** if their joint PDF is of the form

$$f_{\mathbf{X}}(\mathbf{x}) = (2\pi)^{-\frac{1}{2}p} |\mathbf{V}_0|^{-\frac{1}{2}} \exp\{-\tfrac{1}{2}(\mathbf{x} - \xi)' \mathbf{V}_0^{-1} (\mathbf{x} - \xi)\} \quad (1.54)$$

where $E[X_j] = \xi_j$ $(j = 1, \ldots, p)$ i.e. $E[\mathbf{X}] = \xi$, and $\mathbf{V}_0$ is the variance–covariance matrix of $X_1, \ldots, X_p$. The diagonal elements of $\mathbf{V}_0$, $\sigma_{11}, \ldots, \sigma_{pp}$ are the variances of $X_1, \ldots, X_p$ respectively, and the off diagonal element σ_{ij} is the covariance of X_i and X_j, with

$$\sigma_{ij} = \rho_{ij}(\sigma_{ii}\sigma_{jj})^{\frac{1}{2}} \tag{1.55}$$

Of course, $\sqrt{\sigma_{ii}}$ is the standard deviation of X_i; ρ_{ij} is the correlation between X_i and X_j.

For $i = 1, \ldots, p$, $X_i \sim N(\xi_i, \sigma_i^2)$. The quadratic form $(\mathbf{x} - \boldsymbol{\xi})'\mathbf{V}_0^{-1}(\mathbf{x} - \boldsymbol{\xi})$ generalizes the quantity $(x - \xi)^2/\sigma^2 = (x - \xi)\sigma^{-2}(x - \xi)$ which appears in the PDF of the univariate normal distribution. The ellipsoids

$$(\mathbf{x} - \boldsymbol{\xi})'\mathbf{V}_0^{-1}(\mathbf{x} - \boldsymbol{\xi}) = c^2 \tag{1.56}$$

(where c^2 is some constant) define contours of equal probability density ($f_X(x) = \text{const.}$). The probability that $\mathbf{X}$ falls within this contour is $\Pr[\chi_p^2 < c^2]$; the volume inside the ellipsoid is

$$\left\{ \frac{\pi^{\frac{1}{2}p}|\mathbf{V}_0|^{\frac{1}{2}}}{\Gamma(1 + \frac{1}{2}p)} \right\} c^p \tag{1.57}$$

$|\mathbf{V}_0|$ is called the **generalized variance** of $\mathbf{X}$.

1.11 SYSTEMS OF DISTRIBUTIONS

In addition to specific individual families of distributions, of the kind discussed in sections 1.3–1.10, there are systems including a wide variety of families of distributions, but based on some simple defining concept. Among these we will describe first the Pearson system and later the Edgeworth (Gram–Charlier) system. Both are systems of continuous distributions.

1.11.1 Pearson system

This is defined as distributions with PDF $(f(x))$ satisfying the differential equation

$$\frac{d(\log f(x))}{dx} = \frac{-(a+x)}{c_0 + c_1 x + c_2 x^2} \qquad (1.58)$$

This system was developed by Karl Pearson (1857–1936), in the decade 1890–1900.

The values of the parameters c_0, c_1, c_2 determine the shape of the graph of $f(x)$. This can vary quite considerably, and depends on the roots of the quadratic equation

$$c_0 + c_1 x + c_2 x^2 = 0 \qquad (1.59)$$

Among the families (commonly called **Pearson type curves**) are

- $c_0 > 0$; $c_1 = c_2 = 0$ – normal distribution (section 1.4)
- (*Type III*) $c_2 = 0$; $c_1 \neq 0$ – Gamma distribution (section 1.5)
- (*Type I*) Roots of (1.59) real but opposite sign – Beta distribution (section 1.6)
- (*Type VII*) $c_1 = 0$, c_0, $c_2 > 0$ – this includes the t distribution (section 1.7)

These families, we have already encountered, and further details are available for example, in Johnson and Kotz (1970, pp. 9–15).

1.11.2 Edgeworth (Gram–Charlier) distributions

These will be encountered in Chapter 4, and will be described there, but a brief account is available in Johnson and Kotz (1970, pp. 15–22).

1.12 FACTS FROM STATISTICAL METHODOLOGY

This section contains information on more advanced topics of statistical theory. This information is needed for full appreciation of a few sections later in the book, but it is not essential to master (or even read) this material to understand and use the results.

1.12.1 Likelihood

For independent continuous variables $X_1, \ldots, X_n$ possessing PDFs the **likelihood function** is simply the product of their PDFs. Thus the likelihood of $(X_1, \ldots, X_n)$, $L(\mathbf{X})$ is given by

$$L(\mathbf{X}) = f_{X_1}(X_1) f_{X_2}(X_2) \cdots f_{X_n}(X_n)$$

$$= \prod_{i=1}^{n} f_{X_i}(X_i) \tag{1.60}$$

(For discrete variables the $f_{X_i}(X_i)$ would be replaced by the probability function $P_{X_i}(X_i)$ where $P_{X_i}(x_i) = \Pr[X_i = x_i]$.)

1.12.2 Maximum likelihood estimators

If the distributions of $X_1, \ldots, X_n$ depend on values of parameters $\theta_1, \theta_2, \ldots, \theta_s \ (=\boldsymbol{\theta})$ then the likelihood function, L, is a function of $\boldsymbol{\theta}$. The values $\hat{\theta}_1, \hat{\theta}_2, \ldots, \hat{\theta}_s \ (=\hat{\boldsymbol{\theta}})$ that maximize L are called maximum likelihood estimators of $\theta_1, \ldots, \theta_s$ respectively.

1.12.3 Sufficient statistics

If the likelihood function can be expressed entirely in terms of functions of $\mathbf{X}$ ('statistics') $T_1(\mathbf{X}), \ldots, T_h(\mathbf{X})$ the set $(T_1, \ldots, T_h)$

are called sufficient for the parameters in the PDFs of the Xs.

It follows that maximum likelihood estimators of θ must be functions of the sufficient statistics $(T_1, \ldots, T_h) = \mathbf{T}$. Also the joint distribution of the Xs, given values of $\mathbf{T}$ does not depend on θ. This is sometimes described, informally, by the phrase, '$\mathbf{T}$ contains all the information on θ.' If it is not possible to find any function $h(\mathbf{T})$ of $\mathbf{T}$ with expected value zero, except when $\Pr[h(\mathbf{T}) = 0] = 1$, then $\mathbf{T}$ is called a complete sufficient statistic (set).

1.12.4 Minimum variance unbiased estimators

It is also true that among unbiased estimators of a parameter γ an estimator determined by the complete set of sufficient statistics $T_1, \ldots, T_h$ will have the minimum possible variance.

Furthermore, given any unbiased estimator of γ, G say, such a minimum variance unbiased estimator (MVUE) can be obtained by deriving

$$G_0(\mathbf{T}) = E[G|\mathbf{T}] \tag{1.61}$$

i.e. the expected value of G, given $\mathbf{T}$.

This is based on the Blackwell–Rao theorem (Blackwell, 1947; Rao, 1945).

Example

If $X_1, \ldots, X_n$ are independent $N(\xi, \sigma^2)$ random variables then the likelihood function is

$$L(\mathbf{X}) = \prod_{i=1}^{n} \frac{1}{\sigma\sqrt{2\pi}} \exp\left[-\frac{1}{2}\left\{\frac{X_i - \xi}{\sigma}\right\}^2 \right]$$

$$= \frac{1}{(\sigma\sqrt{2\pi})^n} \exp\left[-\frac{1}{2\sigma^2} \sum_{i=1}^{n} (X_i - \xi)^2 \right]$$

$$= \frac{1}{(\sigma\sqrt{2\pi})^n} \exp\left[-\frac{1}{2\sigma^2} \left\{ \sum_{i=1}^{n} (X_i - \bar{X})^2 + n(\bar{X} - \xi)^2 \right\} \right]$$

$$(1.62)$$

The maximum likelihood estimator of ξ is

$$\hat{\xi} = \bar{X}. \tag{1.63 a}$$

The maximum likelihood estimator of σ is

$$\hat{\sigma} = \left\{ n^{-1} \sum_{i=1}^{n} (X_i - \bar{X})^2 \right\}^{\frac{1}{2}} \tag{1.63 b}$$

$(\bar{X}, \hat{\sigma}^2)$ are sufficient statistics for (ξ, σ). They are also, in fact complete sufficient statistic $E[\hat{\xi}] = \xi$, so $\hat{\xi}$ is an unbiased estimator of ξ.

Since $E[\bar{X}|\bar{X}, \hat{\sigma}^2] = \bar{X}$, $\hat{\xi}(=\bar{X})$ is also a minimum variance unbiased estimator of ξ.

But since

$$E[\hat{\sigma}] = \frac{\Gamma(\frac{1}{2}n)}{\Gamma(\frac{1}{2}(n-1))} \left(\frac{2}{n}\right)^{\frac{1}{2}} \sigma \neq \sigma \tag{1.64}$$

it is not an unbiased estimator of σ. However

$$\hat{\sigma} \left(\frac{n}{2}\right)^{\frac{1}{2}} \frac{\Gamma(\frac{1}{2}n-1))}{\Gamma(\frac{1}{2}n)}$$

is not only unbiased, but is a MVUE of σ. (There are more elaborate applications of the Blackwell–Rao theorem in Appendices 2A and 3A.)

1.12.5 Likelihood ratio tests

Suppose we want to choose between hypotheses H_j ($j=0,1$) specifying the values of parameters $\boldsymbol{\theta}=(\theta_1,\ldots,\theta_s)$ to be $\boldsymbol{\theta}_j= (\theta_{1j},\ldots,\theta_{sj})$ ($j=0,1$). The two likelihood functions are

$$L(\mathbf{X}|H_j)= \prod_{i=1}^{n} f_{x_i}(x_i|\boldsymbol{\theta}_j) \quad j=0,1 \tag{1.65}$$

The likelihood ratio test procedure is of the following form

$$\text{If } \frac{L(\mathbf{X}|H_1)}{L(\mathbf{X}|H_0)} <c, \quad \text{choose } H_0$$

$$\text{If } \frac{L(\mathbf{X}|H_1)}{L(\mathbf{X}|H_0)} \geqslant c, \quad \text{choose } H_1 \tag{1.66}$$

The constant c can be chosen arbitrarily. If sufficient information on costs and frequencies with which H_0, H_1 respectively occur is available, it is possible to choose c to produce optimal results in terms of expected cost.

Sometimes c is chosen so as to make the probability of not choosing H_0, when it is, indeed, valid i.e. making an erroneous density equal to some predetermined value, ε, say. The procedure is then termed a 'test of H_0 against the alternative hypothesis H_1, at significance level ε'. H_0 is often termed the 'null' hypothesis, even when there is no special meaning attached to the word 'null' in the particular circumstances.

The significance level, ε, is also called the probability of error of the 'first kind'. The other kind of error, choosing H_0 when the alternative H_1, is valid is called an error of the 'second kind'.

BIBLIOGRAPHY

Blackwell, D. (1947) Conditional expectation and unbiased sequential estimation, *Ann. Math. Statist.*, **18**, 105–110.

Boyles, R.A. (1991) The Taguchi capability index. *Journal of Quality Technology*, **23**(1), 17–23.

Burke, R.J., Davis, R.D. and Kaminsky, F.C. (1991) Responding to statistical terrorism in quality control, *Proc. 47th Ann. Conf. Amer. Soc. Quat. Control*, Rochester Section, March 12.

Chan, L.K., Cheng, C.W., and Spiring, F.A. (1988) A new measure of process capability, *Journal of Quality Technology*, **20**(3), 162–175.

Dovich, R.A. (1991 *a*) in *ASQC Statistical Division Newsletter*, Spring, 5.

Dovich, R.A. (1991 *b*) *Statistical Terrorists II – its not safe yet, C_{pk} is out there*, MS, Ingersoll Cutting Tools Co., Rockford, Illinois.

Elandt, R.C. (1961) The folded normal distribution: two methods of estimating parameters from moments, *Technometrics*, **3**, 551–62.

Gunter, B.H. (1989 *a*) in *Quality Progress*, January, 72.

Gunter, B.H. (1989 *b*) in *Quality Progress*, April.

Johnson, N.L. and Kotz, S. (1970) *Distributions in Statistics: Continuous Univariate Distribution – 1*, Wiley: New York.

Kitska, D. (1991) in *Quality Progress*, March, 8.

McCormick, C. (1989) in *Quality Progress*, April, 9–10.

McCoy, P.F. (1991) in *Quality Progress*, February, 49–55.

Pearn, W.L., Kotz, S. and Johnson, N.L. (1992) Distributional and inferential properties of process capability indices, *J. Qual. Technol.*, **24**, 216–231.

Rao, C.R. (1945) Information and the accuracy attainable in the estimation of statistical parameters. *Bull. Calcutta Math. Soc.*, **37**, 81–91.

Spiring, F.A. (1991) in *Quality Progress*, February, 57–61.

Steenburgh, T. (1991) in *Quality Progress*, January, 90.

2

The basic process capability indices: C_p, C_{pk} and their modifications

2.1 INTRODUCTION

Much of the material presented in this chapter has also appeared in several books on **statistical quality control**, published in the last decade. These books usually provide clear explanations and motivation for the use of the PCIs they describe, and often (though not always) emphasize their limitations, and provide warnings intended to assist recognition of situations wherein extreme caution is needed. In the last few years, also, many issues of *Quality Progress* contain letters to the Editor in which the use of PCIs is criticized (and only occasionally defended), often combined with passionate appeals for their discontinuance on the grounds that their inherent weaknesses guarantee widespread abuse. Some referees for this book have even warned us that we may cause serious damage by further propagating the inadequate and dangerous concepts implicit in use of PCIs.

Our opinion is that we hope (even believe) that we may provide background for rational use of PCIs, based, indeed, on knowledge of their weaknesses. We do *not*, of course, recommend uncritical or exclusive use of these indices. The backlash against PCIs seems to come from two broad, and opposed, sources: (i) a lack of appreciation of statistical theory and its applications; and (ii) a demand for precise statistical analysis, to exclusion of other constraints. Source (i) does not fully comprehend the meaning of the indices, and resents pressure from above to include 'meaningless' numbers in their reports – which may, indeed, contradict the actual state of the process as seen by experienced operators. They regard this as part of the 'tyranny of numbers' which has its roots in the proliferation of statistical (i.e. numerical) information now becoming available, and tending to dominate our existence. Often, this resentment underlies the unpopularity of, and lack of respect for 'statistics' among otherwise enlightened engineers, technicians, sales managers and other professionals.

Source (ii), on the other hand, arises from those who are well aware that assumptions, on which PCIs are based, are often violated in practice. This is also true of many other statistical procedures (such as ANOVA) which, however, are better tolerated, and more widely accepted, albeit somewhat grudgingly.

After these introductory remarks, we now proceed to define and examine the earliest form of PCI, generally denoted by C_p. The background information in Chapter 1 will assist in understanding the basis for many of our results – especially in regard to estimators of PCIs – but they are not essential for general comprehension.

2.2 THE C_p INDEX

Consider a situation wherein there are lower and upper specification limits (LSL, USL respectively) for the value of a measured characteristic X. Values of X outside these limits will be termed 'nonconforming' (NC). An indirect measure of potential ability ('capability') to meet the requirement $(LSL < X < USL)$ is the **process capability index**.

$$C_p = \frac{USL - LSL}{6\sigma} \qquad (2.1\ a)$$

where σ denotes the standard deviation of X.

Clearly, large values of C_p are desirable and small values undesirable (because a large standard deviation is undesirable). The motivation for the multiplier '6' in the denominator seems to be as follows:

If the distribution of X is normal, and if ξ, the expected value of X, is equal to $\frac{1}{2}(LSL + USL)$ – the mid-point of the specification interval – then the expected proportion of NC product is $2\Phi(-d/\sigma)$ where $d = \frac{1}{2}(USL - LSL)$ – the half-length of the specification interval.

From (2.1 a) we see that

$$C_p = \frac{d}{3\sigma} \qquad (2.1\ b)$$

so the expected proportion (p) of NC product (assuming $\xi = \frac{1}{2}(LSL + USL)$) is

$$2\Phi(-3\,C_p) \qquad (2.2)$$

If $C_p = 1$ this expected proportion is 0.27%, which is regarded by some as 'acceptably small'. In fact, it is often required that for acceptance we should have $C_p \leqslant c$ with $c = 1, 1.33,$ or 1.5 corresponding to $USL - LSL = 6\sigma, 8\sigma$ or 9σ.

It is important to note that $C_p = 1$ does *not* guarantee that there will be only 0.27% of NC product. In fact, all that it does guarantee is that, with the assumption of normality and the relevant value of σ, there will *never be less than 0.27%* expected proportion of NC product! (It is *only* when $\xi = \frac{1}{2}(LSL + USL)$ that the expected proportion is as small as

0.27%.) What the value $C_p = 1$ does indicate is that it is *possible* to have the expected proportion of NC product as small as 0.27%, provided the process mean ξ is precisely controlled at $\frac{1}{2}(\text{LSL} + \text{USL})$.

More recently, the view has been put forward that expected proportion of NC product is not (or, perhaps, should not be) the primary motivation in use of PCIs, but rather that loss function considerations should prevail. Although there are some attractions in this attitude, it does not explain the multiplier '6' in the denominator of (2.1 *a*). Also, uncertainty in actual proportion NC is balanced (perhaps *more than* balanced) by uncertainty in knowledge of the true loss function. The form of loss function by far the most favoured is a quadratic function of ξ, though little evidence for this choice has appeared. Also, if one is really interested in a loss function, why not just estimate its average value, and not some unnecessarily complicated function (e.g. reciprocal) of it? Indeed, there would seem to be no need for LSL and USL to appear in the PCI formula at all (except in as far as they may define the loss function). The title and content of a recent paper (Carr (1991)) seem to imply, not only that the original motives underlying definition of PCI have been forgotten, but that they are now returning. Constable and Hobbs (1992) also define 'capable' as referring to percentage of output within specification.

It has even been suggested that $2\Phi(-3C_p)$ itself be used as a PCI, since it is the minimum expected (or the potentially attainable) proportion of NC items *provided* that normality is a valid assumption. Lam and Littig (1992) on the other hand suggest defining a PCI

$$C_{pp} = \tfrac{1}{3}\Phi^{-1}(\tfrac{1}{2}(p+1))$$

and using the estimator

$$\hat{C}_{pp} = \tfrac{1}{3}\Phi^{-1}(\tfrac{1}{2}(\hat{p}+1))$$

based on an estimator of p from the observed values of X. Wierda (1992) suggests using $-\frac{1}{3}\Phi^{-1}(\hat{p})$. Appendix 2.A contains discussion of estimators of p, based on the assumption that the process distribution is normal.

Herman (1989) provides a thought-provoking criticism of the PCI concept, based on engineering considerations. He distinguishes between the 'mechanical industries' (e.g. automotive, wherein C_p first arose) and the 'process industries'. In the former, 'emphasis is on the making of an assembly of parts with little if any measurement error of consequence', while this is not generally true for the process industries.

The σ in the denominator is intended to represent process variability when production is 'in control', but the quantity estimated by

$$S^2 = \frac{1}{n-1} \sum_{j=1}^{n} (X_j - \bar{X})^2$$

(see (2.3) below) is (assuming no bias in measurement error) equal to $\sigma^2 + $ (variance of measurement error).

Herman further notes that if allowance is made for lot-to-lot variability, both σ^2 and the second term have two components – from within-lots and among-lots variation. Again, the σ in the denominator of C_p is intended to refer to *within*-lot process variation. This can be considerably less than the overall standard deviation – σ_{total}, say. Herman suggests that a different index, the 'process performance index' (PPI).

$$P_p = \frac{USL - LSL}{6\,\sigma_{total}}$$

might 'have more value to a customer than C_p'.

Table 2.1 gives values of the *minimum* possible expected proportion $(2\Phi(-3C_p))$ of NC items corresponding to

Table 2.1 Minimum expected proportion of NC items

C_p	2.00	$1\frac{2}{3}$	$1\frac{1}{3}$	1.00	$\frac{2}{3}$	$\frac{1}{3}$
$2\Phi(-3C_p)$	$0.0^6 2\%$	$0.0^4 57\%$	0.0063%	0.27%	4.55%	31.73%
	(=0.002 ppm)	(=0.57 ppm)	(=63 ppm)	(=2700 ppm)	(=45 500 ppm)	(=317 300 ppm)

['ppm' = parts per million]

selected values of C_p, on the assumption of normal variation. One can see why small values of C_p are bad signs! But large values of C_p do not 'guarantee' acceptability, in the absence of information about the value of the process mean.

Montgomery (1985) cites recommended minimum values for C_p, as follows:

- for an existing process, 1.33
- for a new process, 1.50

For characteristics related to essential safety, strength or performance features – for example in manufacturing of bolts used in bridge construction – minimum values of 1.50 for existing, and 1.67 for new processes, are recommended.

Generally, we prefer to emphasize the relations of values of PCIs to expected proportions of NC items, whenever this is possible.

2.3 ESTIMATION OF C_p

The only parameter in (2.1 a) which may need to be estimated is σ, the standard deviation of X. A natural estimator of σ is

$$\hat{\sigma} = \left\{ \frac{1}{n-1} \sum_{j=1}^{n} (X_j - \bar{X})^2 \right\}^{\frac{1}{2}} \tag{2.3}$$

where

$$\bar{X} = \frac{1}{n} \sum_{j=1}^{n} X_j$$

If the distribution of X is normal, then $\hat{\sigma}^2$ is distributed as

$$\frac{1}{n-1} \sigma^2 \times (\text{chi-squared with } (n-1) \text{ degrees of freedom}),$$

$$\tag{2.4 a}$$

or, symbolically

$$\frac{1}{n-1}\sigma^2\chi^2_{n-1} \qquad (2.4\,b)$$

(see section 1.5).

Estimation of σ using random samples from a normal population has been very thoroughly studied. If only

$$C_p^{-1}=\frac{6}{\text{USL}-\text{LSL}}\sigma=\frac{3}{d}\sigma \qquad (2.5)$$

had been used as a PCI, existing theory could have been applied directly to

$$\hat{C}_p^{-1}=\frac{6}{\text{USL}-\text{LSL}}\hat{\sigma}=\frac{3\hat{\sigma}}{d} \qquad (2.6)$$

which is distributed as

$$C_p^{-1}\{\chi_{n-1}/\sqrt{(n-1)}\}$$

Analysis for

$$\hat{C}_p=\frac{\text{USL}-\text{LSL}}{6\hat{\sigma}}=\frac{d}{3\hat{\sigma}}=\frac{\sigma}{\hat{\sigma}}C_p \qquad (2.7)$$

(distributed as $C_p\sqrt{n-1}/\chi_{n-1}$) is a bit more complicated, but not substantially so. Nevertheless, there are several papers studying this topic, including Kane (1986), Chou *et al.* (1990), Chou and Owen (1989) and Li, Owen and Borrego (1990). We give here a succinct statement of results for the distribution of $\hat{C}_p$, assuming X has a normal distribution. Fig. 2.1 (see p. 65) exhibits PDFs of $\hat{C}_p$ for various values of $d/\sigma=3C_p$. From

(2.4) and (2.7), we have

$$\Pr\left[\frac{\hat{C}_p}{C_p} > c\right] = \Pr[\chi^2_{n-1} < (n-1)c^{-2}] \tag{2.8}$$

Since $\Pr[\chi^2_{n-1} \leqslant \chi^2_{n-1,\varepsilon}] = \varepsilon$, we have

$$\Pr\left[\frac{\sigma^2}{n-1}\chi^2_{n-1,\alpha/2} \leqslant \hat{\sigma}^2 \leqslant \frac{\sigma^2}{n-1}\chi^2_{n-1,1-\alpha/2}\right] = 1-\alpha$$

The interval

$$\left(\frac{(n-1)\hat{\sigma}^2}{\chi^2_{n-1,1-\alpha/2}}, \frac{(n-1)\hat{\sigma}^2}{\chi^2_{n-1,\alpha/2}}\right) \tag{2.9 a}$$

is a $100(1-\alpha)\%$ confidence interval for σ^2; and so the interval

$$\left(\frac{6}{\text{USL}-\text{LSL}}\frac{(n-1)^{\frac{1}{2}}}{\chi_{n-1,1-\alpha/2}}\hat{\sigma}, \frac{6}{\text{USL}-\text{LSL}}\frac{(n-1)^{\frac{1}{2}}}{\chi_{n-1,\alpha/2}}\hat{\sigma}\right) \tag{2.9 b}$$

is a $100(1-\alpha)\%$ confidence interval for C_p^{-1}; and

$$\left(\frac{\text{USL}-\text{LSL}}{6\hat{\sigma}}\frac{\chi_{n-1,\alpha/2}}{(n-1)^{\frac{1}{2}}}, \frac{\text{USL}-\text{LSL}}{6\hat{\sigma}}\frac{\chi_{n-1,1-\alpha/2}}{(n-1)^{\frac{1}{2}}}\right)$$

$$\equiv \left(\frac{\chi_{n-1,\alpha/2}}{(n-1)^{\frac{1}{2}}}\hat{C}_p, \frac{\chi_{n-1,1-\alpha/2}}{(n-1)^{\frac{1}{2}}}\hat{C}_p\right) \tag{2.9 c}$$

is a $100(1-\alpha)\%$ confidence interval for C_p. Lower and upper $100(1-\alpha)\%$ confidence limits for C_p are, of course,

$$\frac{\chi_{n-1,\alpha}}{(n-1)^{\frac{1}{2}}}\hat{C}_p \quad \text{and} \quad \frac{\chi_{n-1,1-\alpha}}{(n-1)^{\frac{1}{2}}}\hat{C}_p$$

respectively.

If tables of percentage points of the chi-squared distribution are not available, it is necessary to use approximate formulae. In the absence of such tables, two commonly used approximations are

$$\chi_{v,\alpha} \cong (v - \tfrac{1}{2})^{\frac{1}{2}} + \frac{z_\alpha}{\sqrt{2}} \qquad \text{(Fisher)}$$

and

$$\chi_{v,\alpha} \cong v^{\frac{1}{2}}\left(1 - \frac{2}{9v} + z_\alpha\left(\frac{2}{9v}\right)^{\frac{1}{2}}\right)^{\frac{3}{2}} \qquad \text{(Wilson–Hilferty)}$$

where $\Phi(z_\alpha) = \alpha$. (cf. section 1.4).

With these approximations we would have (from (2.9 c)) the $100(1-\alpha)\%$ confidence interval for C_p:

$$\frac{1}{(n-1)^{\frac{1}{2}}}\hat{C}_p\left(\left(n - \frac{3}{2}\right)^{\frac{1}{2}} - \frac{z_{1-\alpha/2}}{\sqrt{2}}\right),$$

$$\frac{1}{(n-1)^{\frac{1}{2}}}\hat{C}_p\left(\left(n - \frac{3}{2}\right)^{\frac{1}{2}} + \frac{z_{1-\alpha/2}}{\sqrt{2}}\right) \qquad (2.10\,a)$$

(using Fisher) or

$$\hat{C}_p\left(1 - \frac{2}{9(n-1)} - z_{1-\alpha/2}\left(\frac{2}{9(n-1)}\right)^{\frac{1}{2}}\right),$$

$$\hat{C}_p\left(1 - \frac{2}{9(m-1)} + z_{1-\alpha/2}\left(\frac{2}{9(n-1)}\right)^{\frac{1}{2}}\right)^{\frac{3}{2}} \qquad (2.10\,b)$$

(using Wilson–Hilferty)

Heavlin (1988), in an unpublished technical report available from the author, attacks the distribution of S^{-1} directly.

Assuming normality he introduces the approximations

$$E[S^{-1}] \cong \left\{ 1 + \frac{3}{4(n-1)} \right\} \sigma^{-1} \qquad (2.11\,a)$$

$$\text{var}(S^{-1}) \cong \frac{1}{2(n-3)\sigma^2} \qquad (2.11\,b)$$

(cf. section 1.2)
From (2.11 b) we would obtain the approximation

$$\text{var}(\hat{C}_p) \cong \frac{1}{2} \frac{C_p^2}{(n-3)} \qquad (2.11\,c)$$

However, Heavlin (1988), presumably using terms of higher order than n^{-1} in (2.11 b) obtains

$$\text{var}(\hat{C}_p) \cong \frac{C_p^2}{2(n-3)} \left(1 + \frac{6}{n-1} \right) \qquad (2.11\,d)$$

The interval

$$\hat{C}_p \left\{ 1 - \left(\frac{1}{2(n-3)} \left(1 + \frac{6}{n-1} \right) \right)^{\frac{1}{2}} z_{1-\alpha/2} \right\},$$

$$\hat{C}_p \left\{ 1 + \left(\frac{1}{2(n-3)} \left(1 + \frac{6}{n-1} \right) \right)^{\frac{1}{2}} z_{1-\alpha/2} \right\} \qquad (2.12)$$

is suggested as a $100(1-\alpha)\%$ confidence interval for C_p.

Additional generality may be gained by assuming solely that $\hat{\sigma}$ is a statistic, independent of $\bar{X}$, distributed (possibly approximately) as $(\chi_f / \sqrt{f})\sigma$. (The above case corresponds to $f = n - 1$.) This includes situations in which σ is estimated as a multiple of sample range (or of sample mean deviation).

From (2.7), $\hat{C}_{\mathrm{p}}$ is distributed as

$$\frac{\mathrm{USL}-\mathrm{LSL}}{6\sigma}\frac{\sqrt{f}}{\chi_f}=\frac{\sqrt{f}}{\chi_f}C_{\mathrm{p}},\qquad(2.13)$$

so $G=(\hat{C}_{\mathrm{p}}/C_{\mathrm{p}})^2$ is distributed as $f|\chi_f^2|$, and the probability density function of G is

$$f_G(g)=\frac{f^{\frac{1}{2}f}}{2^{\frac{1}{2}f}\Gamma(\frac{1}{2}f)}g^{-\frac{1}{2}f-1}\exp[-\tfrac{1}{2}fg^{-1}]\quad 0<g\qquad(2.14)$$

The rth moment of $\hat{C}_{\mathrm{p}}$ about zero is

$$\mathrm{E}[\hat{C}_{\mathrm{p}}^r]=f^{\frac{1}{2}r}C_{\mathrm{p}}^r\mathrm{E}[\chi_r^{-r}]=f^{\frac{1}{2}r}C_{\mathrm{p}}^r\mathrm{E}[(\chi_f^2)^{-\frac{1}{2}r}]$$
$$=\left(\frac{f}{2}\right)^{\frac{1}{2}r}\frac{\Gamma(\frac{1}{2}(f-r))}{\Gamma(\frac{1}{2}f)}C_{\mathrm{p}}^r.\qquad(2.15)$$

In particular

$$\mathrm{E}[\hat{C}_{\mathrm{p}}]=\left(\frac{f}{2}\right)^{\frac{1}{2}}\frac{\Gamma\left(\dfrac{f-1}{2}\right)}{\Gamma\left(\dfrac{f}{2}\right)}C_{\mathrm{p}}=\frac{1}{b_f}C_{\mathrm{p}}\qquad(2.16a)$$

$$\mathrm{E}[\hat{C}_{\mathrm{p}}^2]=\frac{f}{f-2}C_{\mathrm{p}}^2,\qquad(2.16b)$$

and

$$\mathrm{var}(\hat{C}_{\mathrm{p}})=\left(\frac{f}{f-2}-b_f^{-2}\right)C_{\mathrm{p}}^2\qquad(2.16c)$$

where

$$b_f = \left(\frac{2}{f}\right)^{\frac{1}{2}} \frac{\Gamma(\frac{1}{2}f)}{\Gamma(\frac{1}{2}(f-1))}$$

The estimator

$$\hat{C}'_p = b_f \hat{C}_p \qquad (2.17a)$$

has expected value C_p and so is an *unbiased estimator* of C_p. Its variance is

$$\text{var}(\hat{C}'_p) = \left\{ \frac{fb_f^2}{f-2} - 1 \right\} C_p^2 \qquad (2.17b)$$

Table 2.2 gives a few values of the *unbiasing factor*, b_f. For f greater than 14, an accurate approximate formula is

$$b_f \cong 1 - \frac{3}{4}f^{-1} \qquad (2.18)$$

(cf $(1.19a)$). With this approximation

$$\text{var}(\hat{C}_p) \cong \frac{f(8f+9)}{(f-2)(4f-3)^2} C_p^2 \qquad (2.19a)$$

$$\text{var}(\hat{C}'_p) \cong \frac{8f+9}{16f(f-2)} C_p^2 \qquad (2.19b)$$

Table 2.3 gives numerical expected values and standard deviations (S.D.s) of $\hat{C}_p$ (*not* $\hat{C}'_p$) for the same values of f as in Table 2.2, and $d/\sigma = 2(1)6$. (Note that $d = \frac{1}{2}(\text{USL} - \text{LSL})$). Since $\text{E}[\hat{C}_p]/C_p$ and $\text{var}(\hat{C}_p)/C_p^2$ do not depend on C_p, one could deduce all values in Table 2.3 from those for $C_p = 1$ ($d/\sigma = 3$), for example.

Table 2.2 Values of $b_f = (2/f)^{\frac{1}{2}} \times \Gamma(\tfrac{1}{2}f)/\Gamma(\tfrac{1}{2}(f-1))$

f	4	9	14	19	24	29	34	39	44	49	54	59
b_f	0.798	0.914	0.945	0.960	0.968	0.974	0.978	0.981	0.983	0.985	0.986	0.987

Note that if $n=f+1$, the values of n are 5(5) 60

Table 2.3 Moments of $\hat{C}_p$: E=E[$\hat{C}_p$]; S.D.=S.D. ($\hat{C}_p/C_p$)

$f=n-1$	$d/\sigma\ (=3C_p)$									
	2		3		4		5		6	
	E	S.D.	E	S.D.	E	S.D.	E	S.D.	E	S.D.
4	0.836	0.437	1.253	0.655	1.671	0.874	2.089	1.092	2.507	1.310
9	0.729	0.198	1.094	0.297	1.459	0.396	1.824	0.496	2.189	0.594
14	0.705	0.145	1.058	0.218	1.410	0.291	1.763	0.364	2.116	0.436
19	0.694	0.120	1.042	0.180	1.389	0.240	1.736	0.300	2.083	0.360
24	0.688	0.104	1.033	0.156	1.377	0.209	1.721	0.261	2.065	0.313
29	0.685	0.094	1.027	0.140	1.369	0.187	1.711	0.234	2.054	0.281
34	0.682	0.086	1.023	0.128	1.364	0.171	1.705	0.214	2.046	0.257
39	0.680	0.079	1.020	0.119	1.360	0.159	1.700	0.198	2.039	0.238
44	0.678	0.074	1.017	0.111	1.357	0.148	1.696	0.186	2.035	0.223
49	0.677	0.070	1.016	0.105	1.354	0.140	1.693	0.175	2.031	0.210
54	0.676	0.067	1.014	0.100	1.352	0.133	1.690	0.166	2.028	0.199
59	0.675	0.063	1.013	0.095	1.351	0.127	1.688	0.158	2.026	0.190
$\infty(C_p)$	0.667	–	1.000	–	1.333	–	1.667	–	2.000	–

The population ('true') values of C_p are the expected values for $f = \infty$.

The distribution of $\hat{C}_p$ when the distribution of X is *not* normal will be considered in Chapter 4.

We note that when $f = 24$, for example (with $n = 25$, if $f = n - 1$) the S.D. of $\hat{C}_p$ is very roughly 14–15% of the expected value (i.e. coefficient of variation is 14–15%). This represents a quite considerable amount of dispersion.

As a further example take Montgomery's (1985) suggested minimum value of C_p for 'existing processes', which is 1.33 (see the end of section 2.2). The corresponding value of d/σ is 4. If a sample of size 20 is available, we see (from $f = 19$ in Table 2.3) that if $C_p = 1.33$, the expected value of $\hat{C}_p$ would be 1.389 and its standard deviation is 0.24. Clearly, the results of using $\hat{C}_p$ will be unduly favourable in over 50% of cases – actually we have from (2.8), with $c = 1$,

$$\Pr[\hat{C}_p > 1.33] = \Pr[\chi^2_{19} < 19] = 53.3\%$$

If n is as small as 5 ($f = 4$) and $C_p = 1.33$, then the standard deviation of $\hat{C}_p$ is 0.87 – over half the value of C_p! 'Estimation' in this situation is of very little use, but if n exceeds 50, fairly reasonable accuracy is attained.

Another disturbing factor – possible effects of non-normality – will be the topic of Chapter 4.

Bayesian estimation of C_p has been studied by Bian and Saw (1993).

2.4 THE C_{pk} INDEX

The C_p index does not require knowledge of the value of the process mean, ξ, for its evaluation. As we have already noted, there is a consequent lack of direct relation between C_p and probability of production of NC product. The C_{pk} index was introduced to give the value of ξ some influence on the value of the PCI. This index is defined as

$$C_{pk} = \frac{\min(USL - \xi, \xi - LSL)}{3\sigma} \qquad (2.20\,a)$$

$$= \frac{d - |\xi - \frac{1}{2}(LSL + USL)|}{3\sigma} \text{ (using (1.14\,e))}$$

$$= \left\{ 1 - \frac{|\xi - \frac{1}{2}(LSL + USL)|}{d} \right\} C_p \qquad (2.20\,b)$$

Since $C_p = d/(3\sigma)$, we must have $C_{pk} \leqslant C_p$, with equality if and only if $\xi = \frac{1}{2}(LSL + USL)$. Kane (1986) also defines 'C_{pk}' for a general target value T by replacing $\frac{1}{2}(LSL + USL)$ by T in (2.20\,b). This index shares, and indeed enhances, the feature of C_p, that small values of the index correspond to worse quality. The numerator of C_{pk} is the (signed) distance of ξ from the nearer specification limit.

Some authors have used the erroneous formula

$$\frac{\min(|\xi - LSL|, |USL - \xi|)}{3\sigma} \qquad (2.20\,c)$$

for C_{pk}. This does, indeed, give the same numerical value as formula (2.20\,a), *provided* ξ is in the specification range, LSL to USL. However, if, for example,

$$\xi = LSL - d \quad \text{or} \quad \xi = USL + d$$

formula (2.20\,c) will give the value $d/(3\sigma)$ for C_{pk} – the *same* value as it would for $\xi = \frac{1}{2}(LSL + USL)$, at the middle of the specification range. The correct value, as given by (2.20\,a) is $-d/(3\sigma)$.

We will assume, in the following discussion, that

$$LSL \leqslant \xi \leqslant USL$$

(If ξ were outside the specification range, C_{pk} would be negative, and the process would clearly be inadequate for controlling values of X.)

If the distribution of X is normal, then the expected proportion of NC product is

$$\Phi\left(\frac{LSL-\xi}{\sigma}\right)+\left\{1-\Phi\left(\frac{USL-\xi}{\sigma}\right)\right\} \qquad (2.21\,a)$$

From (2.20 a) we see that, if $\frac{1}{2}(USL+LSL)\leqslant\xi\leqslant USL$ then

$$C_{pk}=\frac{USL-\xi}{3\sigma}$$

and

$$\frac{LSL-\xi}{3\sigma}=\frac{(USL-\xi)-(USL-LSL)}{3\sigma}=C_{pk}-2C_p\leqslant-C_{pk}$$

(since $C_p\geqslant C_{pk}$).

The exact expected proportion of NC product can be expressed in terms of the two PCIs C_p and C_{pk}, as follows:

$$\Phi(-3(2C_p-C_{pk}))+\Phi(-3C_{pk}) \qquad (2.21\,b)$$

Hence the expected proportion of NC product is less than

$$2\Phi(-3C_{pk}) \qquad (2.22\,a)$$

but it is greater than

$$\Phi(-3C_{pk}) \qquad (2.22\,b)$$

The case $LSL\leqslant\xi\leqslant\frac{1}{2}(LSL+USL)$ can be treated similarly.

Although the preceding formulae enable information about expected proportion NC to be expressed in terms of C_{pk} (and

C_p), it can also (and more simply) be expressed in terms of the process mean, ξ, and standard deviation σ.

Porter and Oakland (1990) discuss the relationships between C_{pk} and the probability of obtaining a sample arithmetic mean, based on a random sample of size n, outside control chart limits. Although it is useful to be aware of such relationships, it is much simpler to express them in terms of process mean and standard deviation, than in terms of PCIs. Relationships between C_p and C_{pk} are also discussed by Gensidy (1985) Barnett (1988) and Coleman (1991).

2.4.1 Historical note

It seems that the original way of compensating for the lack of input from ξ in the calculation of C_p, was to use as an *additional* statistic,

$$k = \frac{|\xi - \frac{1}{2}(LSL + USL)|}{d} \tag{2.23}$$

(Kane (1986) defines

$$k = \frac{|\xi - T|}{d}$$

but assumes $T = \frac{1}{2}(LSL + USL)$ in the sequel.)

This was equivalent to using ξ (through k) and σ (through C_p) separately. While it might be said that one might as well use ξ and σ themselves, or their straightforward estimators, $\bar{X}$ and $\hat{\sigma}$, respectively, k and C_p do have the advantage of being dimensionless, and related to the specification limits.

With their combination into the index

$$C_{pk} = (1 - k)C_p \tag{2.24}$$

the reduction to a single index had to be paid for by losing separate information on location (ξ) and dispersion (σ).

Another way of approaching C_{pk} is to introduce the 'lower' and 'upper' PCIs.

$$C_{pkl} = \frac{\xi - LSL}{3\sigma} \qquad (2.25\,a)$$

$$C_{pku} = \frac{USL - \xi}{3\sigma} \qquad (2.25\,b)$$

and to define

$$C_{pk} = \min(C_{pkl}, C_{pku}) \qquad (2.25\,c)$$

Kane (1986) uses the notation C_{pl}, C_{pu} rather than C_{pkl}, C_{pku}. He also introduces indices

$$_T C_{pkl} = \frac{T - LSL}{3\sigma} \quad \text{and} \quad _T C_{pku} = \frac{USL - T}{3\sigma}$$

(our notation).

A similar PPI to P_p (see section 2.2) can be constructed by modifying C_{pk} to

$$P_{pk} = \frac{\min(\xi - LSL, \ USL - \xi)}{3\sigma_{total}}$$

2.5 ESTIMATION OF C_{pk}

A natural estimator of C_{pk} is

$$\hat{C}_{pk} = \frac{d - |\bar{X} - \frac{1}{2}(LSL + USL)|}{3\hat{\sigma}} \qquad (2.26)$$

where $\hat{\sigma}$ is an estimator of σ.

If the distribution of X is normal then not only do we have $\hat{\sigma}$ distributed (perhaps approximately) as $\chi_f \sigma / \sqrt{f}$, but also: (a) $\bar{X}$ is normally distributed with expected value ξ and standard deviation $\sigma / \sqrt{n}$; and (b) $\bar{X}$ and $\hat{\sigma}$ are mutually independent.

From (2.26), using (a) and (b), we find

$$E[\hat{C}_{pk}^r] = \frac{1}{3^r} E[\hat{\sigma}^{-r}] \sum_{j=0}^{r} (-1)^j \binom{r}{j} d^{r-j} E[|\bar{X} - \tfrac{1}{2}(LSL + USL)|^j]$$

$$= \left(\frac{d\sqrt{f}}{3\sigma}\right)^r E[\chi_f^{-r}] \sum_{j=0}^{r} (-1)^j \binom{r}{j}$$

$$\times \left(\frac{\sigma}{d\sqrt{n}}\right)^j E\left[\left| \frac{\sqrt{n}\{\bar{X} - \tfrac{1}{2}(LSL + USL)\}}{\sigma} \right|\right] \qquad (2.27)$$

The statistic $\sqrt{n}|\bar{X} - \tfrac{1}{2}(LSL + USL)|/\sigma$ has a 'folded normal' distribution (see section 1.8), as defined by Leone, Nelson and Nottingham (1961). It is distributed as

$$|U + \delta|$$

where U is a standard normal variable and

$$\delta = \frac{\sqrt{n}}{\sigma}\{\xi - \tfrac{1}{2}(LSL + USL)\} \qquad (2.28)$$

Taking $r = 1, 2$ we obtain

$$E[\hat{C}_{pk}] = \frac{1}{3}\left(\frac{f}{2}\right)^{\frac{1}{2}} \frac{\Gamma\left(\frac{f-1}{2}\right)}{\Gamma\left(\frac{f}{2}\right)} \left[\frac{d}{\sigma} - \left(\frac{2}{\pi n}\right)^{\frac{1}{2}}\right]$$

$$\times \exp\left\{-\frac{n\{\xi-\frac{1}{2}(LSL+USL)\}^2}{2\sigma^2}\right\} - \frac{|\xi-\frac{1}{2}(LSL+USL)|}{\sigma}$$

$$\times\left\{1-2\Phi\left(\frac{-\sqrt{n}\,|\xi-\frac{1}{2}(LSL+USL)|}{\sigma}\right)\right\}\Bigg] \qquad (2.29)$$

and

$$\text{var}(\hat{C}_{pk}) = \frac{f}{9(f-2)}\left[\left(\frac{d}{\sigma}\right)^2 - 2\left(\frac{d}{\sigma}\right)\left[\left(\frac{2}{\pi n}\right)^{\frac{1}{2}}\right.\right.$$

$$\times\exp\left\{-\frac{n[\xi-\frac{1}{2}(LSL+USL)]^2}{2\sigma^2}\right\} + \frac{|\xi-\frac{1}{2}(LSL+USL)|}{\sigma}$$

$$\times\left\{1-2\Phi\left(\frac{-\sqrt{n}\,|\xi-\frac{1}{2}(LSL+USL)|}{\sigma}\right)\right\}\Bigg]$$

$$+\frac{\{\xi-\frac{1}{2}(LSL+USL)\}^2}{\sigma^2} + \frac{1}{n}\Bigg] - \{E[\hat{C}_{pk}]\}^2 \qquad (2.30)$$

(see also Zhang *et al.* (1990))
If we use

$$\hat{\sigma} = \left[\frac{1}{n-1}\sum_{i=1}^{n}(X_i-\bar{X})^2\right]^{\frac{1}{2}},$$

then $f = n-1$. Some numerical values of $E[\hat{C}_{pk}]$ and S.D.$(\hat{C}_{pk})$ are shown in Table 2.4 and corresponding values of C_{pk} are shown in Table 2.5.

Note that $\hat{C}_{pk}$ is a biased estimator of C_{pk}. The bias arises from two sources:

(a) $E[\hat{\sigma}^{-1}] = b_f^{-1}\sigma^{-1} \neq \sigma^{-1}$.

This bias is positive because $b_f < 1$ – see Table 2.2

(b) $E\left[\dfrac{\sqrt{n}\,|\bar{X}-\frac{1}{2}(\text{LSL}+\text{USL})|}{\sigma}\right]-\dfrac{\sqrt{n}\,|\xi-\frac{1}{2}(\text{LSL}+\text{USL})|}{\sigma}\geqslant 0.$

This leads to negative bias, because $\sqrt{n}\,|\bar{X}-\frac{1}{2}(\text{LSL}+\text{USL})|$ has a negative sign in the numerator of $\hat{C}_{\text{pk}}$.

The resultant bias is positive for all cases shown in Table 2.4 for which $\xi\neq\frac{1}{2}(\text{LSL}+\text{USL})=m$. When $\xi=m$, the bias is positive for $n=10$, but becomes negative for larger numbers. Of course, as $n\to\infty$, the bias tends to zero. Table 2.6 exhibits the behaviour of $E[\hat{C}_{\text{pk}}]$, when $d/\sigma=3$ ($C_{\text{p}}=C_{\text{pk}}=1$), as n increases, in some detail.

Studying the values of S.D. $(\hat{C}_{\text{pk}})$ we note that the standard deviation increases as d/σ increases, but decreases as $|\xi-\frac{1}{2}(\text{USL}+\text{LSL})|$ increases. As one would expect, the standard deviation decreases as sample size (n) increases. For interpretation of calculated values of $\hat{C}_{\text{pk}}$ it is especially important to bear in mind the standard deviation (S.D.) of the estimator. It can be seen that a sample size as small as 10 cannot be relied upon to give results of much practical value. For example, when $d/\sigma=3$ ($C_{\text{p}}=1$) the standard deviation of $\hat{C}_{\text{pk}}$ for various values of $\sigma^{-1}|\xi-\frac{1}{2}(\text{LSL}+\text{USL})$ is shown in Table 2.7, together with the true value, C_{pk}.

Even when n is as big as 40 there is still substantial uncertainty in the estimator of C_{pk}, and it is unwise to use $\hat{C}_{\text{pk}}$, as a peremptory guide to action, unless large sample sizes (which may be regarded as uneconomic or even infeasible) are available.

Chou and Owen (1989) utilize formula (2.20 a) as their starting point for deriving the distribution of $\hat{C}_{\text{pk}}$. We have

$$\hat{C}_{\text{pk}}=\min(\hat{C}_{\text{pkl}},\hat{C}_{\text{pku}}) \tag{2.31}$$

where

$$\hat{C}_{\text{pkl}}=\frac{\bar{X}-\text{LSL}}{3\hat{\sigma}} \quad \text{and} \quad \hat{C}_{\text{pku}}=\frac{\text{USL}-\bar{X}}{3\hat{\sigma}}$$

Table 2.4 Moments of C_{pk}: $E = E[\hat{C}_{pk}]$; $S.D. = S.D.(\hat{C}_{pk})$

| | $|\xi - m|/\sigma$ | | | | | | | | | |
| | 0 | | 0.5 | | 1 | | 1.5 | | 2 | |
	E	S.D.	E	S.D.	E	S.D.	E	S.D.	E	S.D.
$n = 10$										
d/σ										
2	0.638	0.188	0.542	0.184	0.365	0.155	0.182	0.129	0.000	0.119
3	1.002	0.282	0.906	0.270	0.729	0.231	0.547	0.191	0.365	0.155
4	1.367	0.378	1.271	0.362	1.094	0.320	0.912	0.275	0.730	0.231
5	1.732	0.476	1.636	0.458	1.459	0.414	1.277	0.367	1.094	0.320
6	2.096	0.574	2.001	0.554	1.824	0.510	1.641	0.462	1.459	0.414
$n = 15$										
d/σ										
2	0.633	0.142	0.527	0.141	0.353	0.118	0.176	0.100	0.000	0.093
3	0.985	0.211	0.880	0.202	0.705	0.173	0.529	0.143	0.353	0.118
4	1.338	0.281	1.232	0.269	1.058	0.237	0.882	0.204	0.705	0.173
5	1.690	0.353	1.585	0.339	1.410	0.305	1.234	0.271	1.058	0.237
6	2.043	0.425	1.938	0.409	1.763	0.375	1.587	0.340	1.411	0.305
$n = 20$										
d/σ										
2	0.633	0.119	0.520	0.119	0.347	0.099	0.174	0.084	0.000	0.079
3	0.980	0.176	0.867	0.169	0.694	0.144	0.521	0.120	0.347	0.099
4	1.327	0.234	1.215	0.224	1.042	0.196	0.868	0.169	0.695	0.143
5	1.674	0.293	1.562	0.281	1.389	0.253	1.215	0.224	1.042	0.196
6	2.022	0.352	1.909	0.330	1.736	0.310	1.563	0.281	1.389	0.252

Table 2.4 (*Cont.*)

					$|\xi-m|/\sigma$					
	0		0.5		1		1.5		2	
	E	S.D.	E	S.D.	E	S.D.	E	S.D.	E	S.D.
$n=25$										
d/σ										
2	0.634	0.105	0.516	0.104	0.344	0.087	0.172	0.074	0.000	0.070
3	0.978	0.154	0.860	0.148	0.688	0.126	0.516	0.105	0.344	0.087
4	1.322	0.205	1.204	0.195	1.032	0.171	0.861	0.148	0.688	0.126
5	1.666	0.256	1.549	0.245	1.377	0.220	1.205	0.195	1.033	0.171
6	2.010	0.307	1.893	0.295	1.721	0.270	1.549	0.245	1.377	0.220
$n=30$										
d/σ										
2	0.635	0.095	0.513	0.094	0.342	0.078	0.171	0.067	0.000	0.063
3	0.977	0.139	0.856	0.133	0.685	0.113	0.513	0.094	0.342	0.079
4	1.319	0.184	1.198	0.175	1.027	0.154	0.856	0.133	0.685	0.113
5	1.662	0.230	1.540	0.220	1.369	0.198	1.198	0.176	1.027	0.154
6	2.004	0.277	1.882	0.265	1.711	0.242	1.540	0.220	1.369	0.198
$n=35$										
d/σ										
2	0.636	0.087	0.511	0.086	0.341	0.072	0.171	0.062	0.000	0.058
3	0.977	0.127	0.852	0.122	0.682	0.103	0.511	0.087	0.341	0.072
4	1.318	0.169	1.193	0.161	1.023	0.141	0.852	0.122	0.682	0.103
5	1.659	0.211	1.534	0.201	1.364	0.181	1.193	0.161	1.023	0.141
6	2.000	0.253	1.875	0.242	1.705	0.222	1.534	0.201	1.364	0.181

n = 40

d/σ										
2	0.637	0.081	0.500	0.080	0.340	0.067	0.170	0.058	0.000	0.054
3	0.977	0.119	0.850	0.113	0.680	0.096	0.510	0.080	0.340	0.067
4	1.317	0.157	1.190	0.149	1.020	0.131	0.850	0.112	0.680	0.096
5	1.657	0.196	1.530	0.186	1.360	0.168	1.190	0.149	1.020	0.131
6	1.991	0.235	1.870	0.225	1.700	0.200	1.530	0.186	1.360	0.168

n = 45

d/σ										
2	0.638	0.076	0.509	0.075	0.339	0.063	0.170	0.054	0.000	0.051
3	0.977	0.111	0.848	0.106	0.678	0.090	0.509	0.075	0.339	0.063
4	1.316	0.147	1.187	0.139	1.018	0.122	0.848	0.106	0.678	0.090
5	1.655	0.184	1.526	0.175	1.357	0.157	1.187	0.139	1.017	0.122
6	1.995	0.220	1.865	0.210	1.696	0.192	1.526	0.175	1.357	0.157

n = 50

d/σ										
2	0.639	0.072	0.508	0.071	0.339	0.060	0.169	0.051	0.000	0.048
3	0.977	0.105	0.846	0.100	0.677	0.085	0.508	0.071	0.339	0.060
4	1.316	0.139	1.185	0.132	1.016	0.116	0.846	0.100	0.677	0.085
5	1.655	0.174	1.524	0.165	1.354	0.148	1.185	0.132	1.016	0.116
6	1.993	0.208	1.862	0.198	1.693	0.182	1.523	0.165	1.354	0.148

Table 2.4 (*Cont.*)

					$|\xi - m|/\sigma$					
	0		0.5		1		1.5		2	
	E	S.D.	E	S.D.	E	S.D.	E	S.D.	E	S.D.
$n=55$										
d/σ										
2	0.640	0.069	0.507	0.068	0.338	0.057	0.169	0.049	0.000	0.046
3	0.978	0.100	0.845	0.095	0.676	0.081	0.507	0.068	0.338	0.057
4	1.316	0.132	1.183	0.125	1.014	0.110	0.845	0.095	0.676	0.081
5	1.654	0.165	1.521	0.156	1.352	0.141	1.183	0.125	1.014	0.110
6	1.992	0.198	1.859	0.188	1.690	0.172	1.521	0.156	1.352	0.141
$n=60$										
d/σ										
2	0.641	0.066	0.506	0.065	0.338	0.054	0.169	0.047	0.000	0.044
3	0.978	0.096	0.844	0.091	0.675	0.077	0.506	0.065	0.338	0.054
4	1.316	0.126	1.182	0.119	1.013	0.105	0.844	0.091	0.675	0.077
5	1.653	0.157	1.519	0.149	1.351	0.134	1.182	0.119	1.013	0.105
6	1.991	0.189	1.857	0.180	1.688	0.164	1.519	0.149	1.351	0.134

Table 2.5 Values of C_{pk}

$\dfrac{d}{\sigma} \Big\backslash \dfrac{|\xi-m|}{\sigma}$	0.0	0.5	1.0	1.5	2.0
2	$\frac{2}{3}$	$\frac{1}{2}$	$\frac{1}{3}$	$\frac{1}{6}$	0
3	1	$\frac{5}{6}$	$\frac{2}{3}$	$\frac{1}{2}$	$\frac{1}{3}$
4	$1\frac{1}{3}$	$1\frac{1}{6}$	1	$\frac{5}{6}$	$\frac{2}{3}$
5	$1\frac{2}{3}$	$1\frac{1}{2}$	$1\frac{1}{3}$	$1\frac{1}{6}$	1
6	2	$1\frac{5}{6}$	$1\frac{2}{3}$	$1\frac{1}{2}$	$1\frac{1}{3}$

$m = \frac{1}{2}(\text{USL} + \text{LSL})$

Table 2.6 Values of $E[\hat{C}_{pk}]$ for $\xi = m$ and $d/\sigma = 3$ corresponding to $C_p = C_{pk} = 1$ for a series of increasing values of n

Sample size n	$E[\hat{C}_{pk}]$
10	1.002
20	0.980
30	0.977
60	0.978
80	0.980
100	0.981
200	0.985
400	0.989
600	0.990
2200	0.995
3200	0.996
5400	0.997
10800	0.998
30500	0.999
79500	1.000

are natural estimators of C_{pkl} and C_{pku} respectively. The distributions of $3\sqrt{n} \times \hat{C}_{pkl}$ and $3\sqrt{n} \times \hat{C}_{pku}$ are noncentral t with f degrees of freedom and noncentrality parameters $\sqrt{n}\,(\xi - \text{LSL})/\sigma$ and $\sqrt{n}\,(\text{USL} - \xi)/\sigma$ respectively. Chou and

Table 2.7 Variability of $\hat{C}_{pk}(C_p=1)$

$|\xi-\frac{1}{2}(\text{LSL}+\text{USL})|/\sigma$	0.0	0.5	1.0	1.5	2.0
C_{pk}	1.00	0.83	0.67	0.50	0.33
S.D. $(\hat{C}_{pk})(n=10)$	0.28	0.27	0.23	0.19	0.155
S.D. $(\hat{C}_{pk})(n=40)$	0.12	0.11	0.095	0.08	0.07

Owen (1989) use Owen's (1965) formulae for the joint distribution of these two dependent noncentral t variables to derive the distribution of $\hat{C}_{pk}$. The formula is too complicated to give here, but it is the basis for a computer program presented by Guirguis and Rodriguez (1992). This paper also contains graphs of the probability density function $\hat{C}_{pk}$ for selected values of the parameters, and sample sizes of 30 and 100.

The following approach may be of interest, and provides some insight into the structure of the distribution of $\hat{C}_{pk}$.

The inequality

$$\hat{C}_{pk}<c$$

is equivalent to

$$d-|\bar{X}-m|<3c\hat{\sigma}$$

i.e.

$$|\bar{X}-m|+3c\hat{\sigma}>d \tag{2.32}$$

Under the stated assumptions, $\sqrt{n}\,|\bar{X}-m|$ is distributed as $\chi_1\sigma$, $(n-1)^{\frac{1}{2}}\hat{\sigma}$ is distributed as $\chi_{n-1}\sigma$ and these two statistics are mutually independent (see end of section 1.5).

Hence

$$\Pr[\hat{C}_{pk}<c]=\Pr\left[\frac{1}{\sqrt{n}}\chi_1+\frac{3c}{(n-1)^{\frac{1}{2}}}\chi_{n-1}>\frac{d}{\sigma}\right] \tag{2.33}$$

where χ_1 and χ_{n-1} are mutually independent.

In (2.33), we see that, as $n \to \infty$, the term $\chi_1 / \sqrt{n}$ becomes negligible compared with $3c\chi_{n-1}/(n-1)^{\frac{1}{2}}$ (which tends to $3c$), and the distribution of $\hat{C}_{pk}$ tends to that of a multiple of a 'reciprocal χ' – in fact, as $n \to \infty$ $\hat{C}_{pk} \cong \hat{C}_p$ and so has the approximate distribution (2.13) with $f = n - 1$.

Figures 2.2–5 show the shape factors ($\sqrt{\beta_1}$ and β_2 – see section 1.1) for the distribution of $\hat{C}_{pk}$. In each figure, the dotted lines represent the locus of points ($\sqrt{\beta_1}, \beta_2$) for the distribution of $\hat{C}_{pk}$ for fixed $d^* = d/\sigma$ (Figs 2.2 and 2.4) with $\delta = |\xi - m|/\sigma$ varying, or for fixed δ (Figs 2.3 and 2.5) with $d^* = d/\sigma$ varying. Figures 2.2 and 2.3 are for sample size (n) equal to 30; Figs 2.4 and 2.5 are for sample size 100. The other lines on the figures represent loci for other standard distributions, to assist in assessing the shape of the $\hat{C}_{pk}$ distribution relative to well-established distributions.

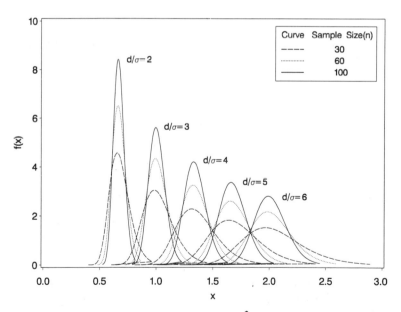

Fig. 2.1 Probability density functions of $\hat{C}_p$ for various sample sizes and values of d/σ.

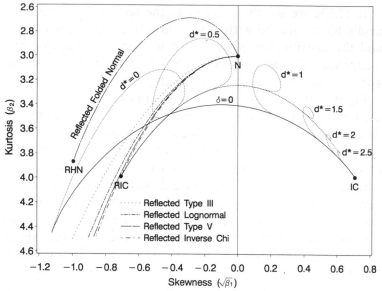

Fig. 2.2 Moment ratios of $\hat{C}_{\mathrm{pk}}$ for $n = 30$ and levels of $d^* = d/\sigma$. Labelled points identify special distributions: N (normal), IC (inverse chi with df = 29), RIC (reflected inverse chi with df = 29), RHN (reflected half normal).

Comparison of Figs 2.2–3 with Figs 2.4–5 shows a greater concentration of $(\sqrt{\beta_1}, \beta_2)$ values near the normal point $(N \equiv (0, 3))$ for the larger sample size $(n = 100)$.

Bissell (1990) uses the modified estimator

$$\hat{C}'_{\mathrm{pk}} = \begin{cases} \dfrac{\mathrm{USL} - \bar{X}}{3\hat{\sigma}} & \text{for } \xi \geqslant \dfrac{\mathrm{LSL} + \mathrm{USL}}{2} \\[3mm] \dfrac{\bar{X} - \mathrm{LSL}}{3\hat{\sigma}} & \text{for } \xi \leqslant \dfrac{\mathrm{LSL} + \mathrm{USL}}{2} \end{cases} \tag{2.34}$$

This differs from C_{pk} only in the use of ξ in place of $\bar{X}$ in the inequalities defining which expression is to be used. Note that $\hat{C}'_{\mathrm{pk}}$ cannot be calculated unless ξ is known. However, if ξ is known, one would not need to estimate it

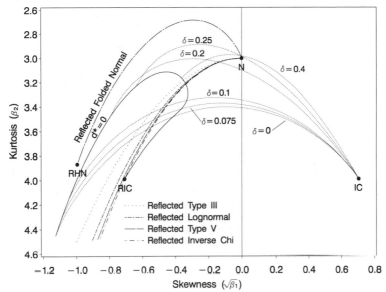

Fig. 2.3 Moment ratios of $\hat{C}_{pk}$ for $n = 30$ and levels of $\delta = |\xi - T|/\sigma$. Labelled points identify special distributions: N (normal), IC (inverse chi with df = 29), RIC (reflected inverse chi with df = 99), RHN (reflected half normal).

and could simply use

$$\hat{C}_{pk}^* = \min(\hat{C}_{pkl}^*, \hat{C}_{pku}^*) \tag{2.35}$$

where

$$\hat{C}_{pku}^* = \frac{\text{USL} - \xi}{3\hat{\sigma}} \quad \text{and} \quad \hat{C}_{pkl}^* = \frac{\xi - \text{LSL}}{3\hat{\sigma}}$$

The distribution of $\hat{C}_{pk}'$, as defined by (2.34), is that of

$\dfrac{1}{3\sqrt{n}} \times$ (noncentral t with f degrees of freedom and noncentrality parameter $\sqrt{n}\{d - |\xi - \frac{1}{2}(\text{LSL} + \text{USL})|\}/\sigma$)

(see section 1.7.)

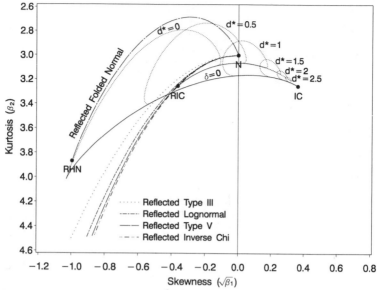

Fig. 2.4 Moment ratios of $\hat{C}_{pk}$ for $n = 100$ and levels of $d^* = d/\sigma$. Labelled points identify special distributions: N (normal), IC (inverse chi with df = 99), RIC (reflected inverse chi with df = 99), RHN (reflected half normal).

Although $\hat{C}'_{pk}$ is not of practical use, it will not differ greatly from $\hat{C}_{pk}$, *except* when ξ is close to $\frac{1}{2}(LSL + USL)$. This is because, the probability that $\bar{X} - \frac{1}{2}(LSL + USL)$ has the same sign as $\xi - \frac{1}{2}(LSL + USL)$ is

$$1 - \Phi\left(- \frac{\sqrt{n} \, |\xi - \frac{1}{2}(LSL + USL)|}{\sigma} \right)$$

and this will be close to 1 for n large, unless $\xi = \frac{1}{2}(LSL + USL)$.

As the distribution (noncentral t) of $\hat{C}'_{pk}$ is a well-established one, it can be used as an approximation to that of $\hat{C}_{pk}$ under suitable conditions – namely that ξ differs from $\frac{1}{2}(LSL + USL)$ by a substantial amount, and n is not too small.

Construction of a confidence interval for C_{pk} is more

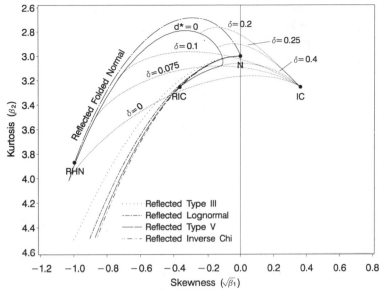

Fig. 2.5 Moment ratios of $\hat{C}_{pk}$ for $n = 100$ and levels of $\delta = |\xi - T|\sigma$. Labelled points identify special distributions: N (normal), IC (inverse chi with df = 99), RIC (reflected inverse chi with df = 99), RHN (reflected half normal).

difficult, because two parameters (ξ, σ) are involved. It is tempting to obtain separate confidence intervals for the two parameters, and then construct a confidence interval for C_{pk}, with limits equal to the least and gretest possible values for C_{pk} corresponding to pairs of values (ξ, σ) within the rectangular region defined by the separate confidence intervals. Although this may give a useful indication of the accuracy of estimation of C_{pk} it is necessary to remember the following two points.

(a) If each of the separate intervals has confidence coefficient $100(1 - \alpha)\%$ the confidence coefficient of the derived interval for C_{pk} will *not* in general be $100(1 - \alpha)\%$. For example, if the statistics used in calculating the separate intervals were mutually independent, the probability that each will include its appropriate parameter value would not be $100(1-\alpha)\%$, but $100(1-\alpha)^2\%$. This would mean, for example, with $\alpha=0.05$, that we would have 90.25%, rather than 95% confidence coefficient. Generally, the combined probability will be *less* than the confidence coefficient of each component confidence interval.

(b) Another effect, arising from the fact that only *part* of the rectangular region contributes to the interval for C_{pk}, and that other pairs of values (ξ, σ) could give C_{pk} values in the same interval, tends to counterbalance effect (a).

In the case of C_{pk}, the $100(1-\alpha)\%$ separate intervals for ξ and σ are

$$\bar{X}-t_{f,\,1-\alpha/2}\hat{\sigma} \leqslant \xi \leqslant \bar{X}+t_{f,\,1-\alpha/2}\hat{\sigma} \qquad (2.36\,a)$$

$$\frac{\hat{\sigma}\sqrt{f}}{\chi_{f,\,1-\alpha/2}} \leqslant \sigma \leqslant \frac{\hat{\sigma}\sqrt{f}}{\chi_{f,\,\alpha/2}} \qquad (2.36\,b)$$

Although $\bar{X}$ and $\hat{\sigma}$ are mutually independent, the events defined by (2.36 a) and (2.36 b) are not independent. However, it is still true that the probability that the interval for C_{pk}, calculated as described above, really includes the true value will be less than $100(1-\alpha)\%$. The extent of the counterbalancing effect of (b) is difficult to ascertain, but it is not likely to be comparable to the effect of (a).

Zhang *et al.* (1990), Kushler and Hurley (1992) and Nagata and Nagaharta (1993) provide a thorough treatment of construction of confidence intervals for C_{pk} which will be relatively simple to calculate.

Heavlin (1988) in the report already referred to, suggests

$$\left(\hat{C}_{pk}-z_{1-\alpha/2}\left\{\frac{n-1}{9n(n-3)}+\hat{C}_{pk}^2\frac{1}{2(n-3)}\left(1+\frac{6}{n-1}\right)\right\}^{\frac{1}{2}},\right.$$

$$\left.\hat{C}_{pk}+z_{1-\alpha/2}\left\{\frac{n-1}{9n(n-3)}+\hat{C}_{pk}^2\frac{1}{2(n-3)}\left(1+\frac{6}{n-1}\right)\right\}^{\frac{1}{2}}\right) \qquad (2.37)$$

as a $100(1-\alpha)\%$ confidence interval for C_{pk}.

Chou, Owen and Borrego (1990) provided tables of 'approximate 95% lower confidence limits' for C_{pk} (*inter alia*) under the assumption of normality. A few of their values are shown below in Table 2.8.

Table 2.8 Approximate 95% lower confidence limits for C_{pk}

$\hat{C}_{pk}$	$n=30$	$n=50$	$n=75$
1.0	0.72	0.79	0.83
1.1	0.80	0.87	0.91
1.5	1.12	1.21	1.26
1.667	1.25	1.35	1.40

Franklin and Wasserman (1992) carried out simulations to assess the properties, of these limits, and discovered that they are conservative. They found that the actual coverage for the limits in Table 2.8, is about 96–7%. Guirguis and Rodriguez (1992, p. 238) explain why these approximate limits are conservative.

On the other hand, Franklin and Wasserman's studies showed that the formula

$$\hat{C}_{pk} - z_{1-\alpha}\left(\frac{1}{9n} + \frac{\hat{C}_{pk}^2}{2(n-1)}\right)^{\frac{1}{2}} \qquad (2.38)\ (cf\ (2.37))$$

produces (for $n \geqslant 30$) remarkably accurate lower $100(1-\alpha)\%$ confidence limits for C_{pk}.

Nagata and Nagahata (1993) suggest modifying (2.38) by adding $1/(30\sqrt{n})$. They obtain very good results from simulation experiments for the two-sided confidence intervals with this modification. The parameter values employed were $C_{pk} = 0.4(0.3)2.5$; $n = 10(10)50$, 100; $1-\alpha = 0.90$, 0.95. The actual coverage was never greater than 0.902 for $1-\alpha = 0.90$, 0.954 for $1-\alpha = 0.95$; and never less than $1-\alpha$.

Kushler and Hurley (1991) suggest the simpler formula

$$\hat{C}_{pk}\left[\frac{1-z_{1-\alpha}}{(2n-2)^{\frac{1}{2}}}\right] \tag{2.39}$$

and Dovich (1992) reports that the corresponding approximate $100(1-\alpha)\%$ formula for confidence interval limits:–

$$\hat{C}_{pk}\left[\frac{1\mp z_{1-\alpha/2}}{(2n-2)^{\frac{1}{2}}}\right] \tag{2.40}$$

gives good results.

*The interested reader may find the following brief discussion illuminating.

Difficulties in the construction of confidence intervals for C_{pk} arise from the rather complicated way in which the parameters ξ and σ appear in the expression for C_{pk}.

We first consider the much simpler problem arising when the value of σ is known. We then calculate

$$C_{pk}^* = \frac{\min(\bar{X}-\text{LSL}, \text{USL}-\bar{X})}{3\sigma}$$

as our estimate of C_{pk}.

The $100(1-\alpha)\%$ confidence interval for ξ,

$$\underline{\xi}(\bar{X}) \leqslant \xi \leqslant \bar{\xi}(\bar{X}) \tag{2.41}$$

with $\underline{\xi}(\bar{X})=\bar{X}-z_{1-\alpha/2}\sigma/\sqrt{n}$, $\bar{\xi}(\bar{X})=\bar{X}+z_{1-\alpha/2}\sigma/\sqrt{n}$ can be used as a basis for constructing a confidence region for

$$C_{pk} = \frac{\min(\xi-\text{LSL}, \text{USL}-\xi)}{3\sigma} \tag{2.42}$$

The confidence region is obtained by replacing the numerator by the range of values it can take for (2.41). A little care is

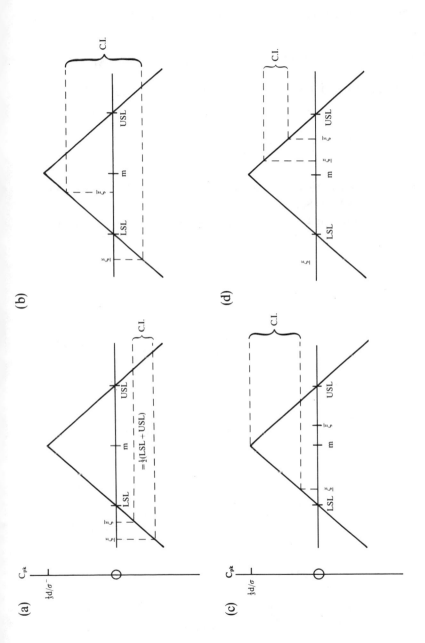

Fig. 2.6 Confidence intervals for C_{pk} (σ known).

needed; this range is *not*, in general, just the interval with end points

$$\min(\underline{\xi}(\bar{X}) - \text{LSL}, \text{USL} - \underline{\xi}(\bar{X})); \min(\bar{\xi}(\bar{X}) - \text{LSL}, \text{USL} - \bar{\xi}(\bar{X})).$$

For example, if $\text{LSL} < \underline{\xi}(\bar{X}) < \frac{1}{2}(\text{LSL} + \text{USL})$ and $\bar{\xi}(\bar{X}) > \text{USL}$ (see Fig. 2.6 c) then the confidence interval for C_{pk} is

$$\frac{1}{3\sigma}(\text{USL} - \bar{\xi}(\bar{X})), \frac{1}{3\sigma} \cdot \frac{1}{2}(\text{USL} - \text{LSL}) = \frac{d}{3\sigma} \qquad (2.43)$$

Some possibilities are shown in Table 2.9 (with $m = \frac{1}{2}(\text{LSL} + \text{USL})$). (See also Figs. 2.6 a–d.)

When σ is not known, but is estimated by S, the problem is more complicated. If we replace the values $\underline{\xi}(\bar{X}), \bar{\xi}(\bar{X})$ by

$$\frac{\bar{X} - t_{n-1,\alpha/2}S}{\sqrt{n}} \quad \text{and} \quad \frac{\bar{X} + t_{n-1,1-\alpha/2}S}{\sqrt{n}}$$

((2.36 a) with $f = n - 1$) we would still get $100(1 - \alpha)\%$ confidence limits from Table 2.9, but computation of these limits would need the correct value of σ to evaluate the multiplier $1/(3\sigma)$.

Table 2.9 Confidence interval for C_{pk} (σ known)

Value of		Confidence limits ($\times 3\sigma$)
$\underline{\xi}(\bar{X})$	$\bar{\xi}(\bar{X})$	
$\leqslant m$	$\leqslant m$	$(\underline{\xi}(\bar{X}) - \text{LSL}, \bar{\xi}(\bar{X}) - \text{LSL})$
$\leqslant m$	$> m$	$(\min(\underline{\xi}(\bar{X}) - \text{LSL}, \text{USL} - \bar{\xi}(\bar{X})), \frac{1}{2}(\text{USL} - \text{LSL}))$
$> m$	$> m$	$(\text{USL} - \bar{\xi}(\bar{X}), \text{USL} - \underline{\xi}(\bar{X}))$

Bayesian estimation of C_{pk} has been studied by Bian and Saw (1993).

2.6 COMPARISON OF CAPABILITIES

PCIs are measures of 'capability' of a process. They may also be used to compare capabilities of two (or more) processes. In particular, if process variation is normal, the index

$$C_{\text{pl}} = \frac{1}{3} \frac{\xi - \text{LSL}}{\sigma}$$

is directly linked to the expected proportion of output which is below the LSL, by the formula $\Phi(-3C_{\text{pl}})$. Comparison of estimators $\hat{C}_{\text{pl1}}$, $\hat{C}_{\text{pl2}}$ of C_{pl} values $(C_{\text{pl1}}, C_{\text{pl2}})$ for two processes is therefore equivalent to comparing proportions of output for the two processes with values falling below their respective LSLs. With sufficient data, of course, this could be effected by the simple procedure of comparing observed proportions of values falling below these limits. However, provided the assumption of normality is valid, the use of $\hat{C}_{\text{pl1}}$, $\hat{C}_{\text{pl2}}$ might be expected to give a more powerful test procedure.

Chou and Owen (1991) have developed such tests, which can be regarded as a test of the hypothesis $C_{\text{pl1}} = C_{\text{pl2}}$ against either two-sided $(C_{\text{pl1}} \neq C_{\text{pl2}})$ or one-sided $(C_{\text{pl1}} < C_{\text{pl2}}$, or $C_{\text{pl1}} > C_{\text{pl2}})$ alternative hypotheses, for the case when the sample sizes from which the estimators $\hat{C}_{\text{pl1}}$, $\hat{C}_{\text{pl2}}$ are calculated are the same – n, say. In an obvious notation the statistics $T_1 = 3\sqrt{n}\hat{C}_{\text{pl1}}$, $T_2 = 3\sqrt{n}\hat{C}_{\text{pl2}}$ are independent noncentral t variables with $(n-1)$ degrees of freedom, and noncentrality parameters $\sqrt{n}(\xi_1 - \text{LSL}_1)/\sigma_1$, $\sqrt{n}(\xi_2 - \text{LSL}_2)/\sigma_2$ respectively. Applying a generalized likelihood ratio test technique (section 1.12), Chou and Owen obtain the test statistic

$$U = [\{T_1^2 + 2(n-1)\}\{T_2^2 + 2(n-1)\}]^{\frac{1}{2}} - T_1 T_2 \qquad (2.44)$$

To establish approximate significance limits for U they carried out simulation experiments as a result of which they suggest the approximation, for the distribution of U, when H_0 is valid:

'$a_n \log\{\frac{1}{2}U/(n-1)\}$ approximately distributed as chi-squared with v_n degrees of freedom.' Table 2.10 gives values of a_n and v_n, estimated by Chou and Owen from their simulations.

Table 2.10 Values of a_n and v_n

n	a_n	v_n
5	8.98	1.28
10	19.05	1.22
15	26.47	1.07
20	36.58	1.06

The progression of values is rather irregular, possibly reflecting random variation from the simulations (though there were 10 000 samples in each case). [If the values of a_{10} and a_{15} were about 18.2 and 27.3 respectively, progression would be smooth.]

2.7 CORRELATED OBSERVATIONS

The above theory based on mutually independent observations of X can be extended to cases in which the observations of the process characteristic are correlated, using results presented by Yang and Hancock (1990). They show that

$$E[S^2] = (1 - \bar{\rho})\sigma^2$$

where $\bar{\rho}$ = average of the $\frac{1}{2}n(n-1)$ correlations between pairs of Xs. (See Appendix 2.A.)

This implies that in the estimator

$$\hat{C}_p = \frac{1}{3}\frac{d}{S}$$

of C_p, the denominator will tend to underestimate σ and so $\hat{C}_p$ will tend to overestimate C_p. This reinforces the effect of the bias in $1/S$ as an estimate of $1/\sigma$, which also tends to produce overestimation of C_p (see Table 2.3).

There will be a similar effect for C_{pk}. We warn the reader, however, that in this case we cannot rely on independence of $\bar{X}$ and S.

Correlation can arise naturally when σ^2 is estimated from data classified into several subgroups. Suppose that we have k subgroups, each containing m items and denote by X_{uv} the value of vth item in the uth subgroup. If the model

$$X_{uv} = Y_u + Z_{uv} \quad u = 1, \ldots, k; v = 1, \ldots, m \tag{2.45}$$

applies with all Ys and Z mutually independent random variables, each with expected value zero, and with

$$\text{var}(Y_u) = \sigma_Y^2 \quad \text{var}(Z_{uv}) = \sigma_Z^2$$

then we have $n = km$,

$$\sigma^2 = \sigma_Y^2 + \sigma_Z^2$$

and correlation between

$$X_{uv} \text{ and } X_{u'v'} = \begin{cases} 0 & \text{if } u \neq u' \\ \dfrac{\sigma_Y^2}{(\sigma_Y^2 + \sigma_Z^2)} & \text{if } u = u' \text{ (and } v \neq v') \end{cases}$$

so that

$$\begin{aligned} \bar{\rho} &= \frac{1}{n(n-1)} km(m-1) \frac{\sigma_Y^2}{(\sigma_Y^2 + \sigma_Z^2)} \\ &= \frac{km(m-1)}{n(n-1)} \frac{\sigma_Y^2}{\sigma^2} \\ &= \frac{m-1}{n-1} \frac{\sigma_Y^2}{\sigma^2} \end{aligned}$$

In the case of equal correlation we have $\rho_{ij} = \bar{\rho}$ for all $i, j (i \neq j)$. In this case S^2 is distributed as $(1 - \bar{\rho})\chi^2_{n-1}\sigma^2/(n-1)$.

2.8 PCIs FOR ATTRIBUTES

An interesting proposal for PCI for attributes (characteristics taking values 0 and 1 only – e.g., conforming (C) or (NC)), was made by Marcucci and Beazley (1988). This has not as yet received attention in the literature.

We suppose that an upper limit ω, for proportion of NC product is specified. It is proposed to use the **odds ratio**

$$R = \frac{\omega(1 - \omega_1)}{(1 - \omega)\omega_1} \tag{2.46}$$

(see, e.g., Fleiss (1973, pp. 61 *et seq.*) or Elandt-Johnson and Johnson (1980, p. 37)) as a PCI, where ω is the actual proportion of NC product. The ratio $\omega(1 - \omega)^{-1}$ measures the odds for the characteristic being NC, and $\omega_1(1 - \omega_1)^{-1}$ is the **maximally acceptable odds**. Thus $R = 1$ corresponds to product which is just acceptable, while $R = 2$ reflects an unfortunate situation, in which the odds for getting a NC item are twice the maximum acceptable odds. Unlike C_p, C_{pk}, etc., large values of this PCI indicate bad situations.

Marcucci and Beazley (1988) study the properties of estimators of R. Given a random sample of size n, with X items found to be NC, the estimator

$$\hat{R} = \frac{X + \frac{1}{2}}{n - X + \frac{1}{2}} \frac{1 - \omega_1}{\omega_1} \tag{2.47}$$

is well-known in statistical practice. The $\frac{1}{2}$s are so-called 'Sheppard corrections' for continuity. Unfortunately, the variance of $\hat{R}$ is large, and so is the (positive) bias, especially in small samples. Marcucci and Beazley (1988), therefore, suggest 'shrinking' $\hat{R}$ towards the value 1.0, to produce a PCI estimator

$$\hat{C}_a = k\hat{R} + (1-k) \tag{2.48}$$

for some k. (The subscript 'a' stands for 'attribute.')

They try to choose k to minimize the mean square error $E[(\hat{C}_a - R)^2]$, and suggest taking

$$k = \frac{1-\hat{R}^2}{\hat{\sigma}_R^2 + 1 - \hat{R}^2} \tag{2.49}$$

where

$$\hat{\sigma}_{R^2} = \hat{R}^2 \{(X + \tfrac{1}{2})^{-1} + (n - X + \tfrac{1}{2})^{-1}\}$$

APPENDIX 2.A

If $X_1, X_2, \ldots, X_n$ are independent normal variables with common expected value ξ and standard deviation σ then

$$1 - p = \Pr[\text{LSL} < X_i < \text{USL}] = \Phi\left(\frac{\text{USL} - \xi}{\sigma}\right) - \Phi\left(\frac{\text{LSL} - \xi}{\sigma}\right) \tag{2.50}$$

A natural estimator of p (the expected proportion of NC items) is

$$\hat{p} = 1 - \Phi\left(\frac{\text{USL} - \bar{X}}{S}\right) + \Phi\left(\frac{\text{LSL} - \bar{X}}{S}\right) \tag{2.51}$$

This estimator is biased. A minimum variance unbiased estimator of p is derived below. It is quite complicated in form and for most occasions (2.51) will be more convenient, and also quite adequate.

We will apply the Blackwell–Rao theorem (section 1.12.4). The statistics $(\bar{X}, S)$ are sufficient for ξ and σ, and $(1-Y)$, where

$$Y = \begin{cases} 1 & \text{if } \text{LSL} < X_1 < \text{USL} \\ 0 & \text{otherwise} \end{cases}$$

is an unbiased estimator of p. Hence the conditional expected value of $1 - Y$, given $(\bar{X}, S)$, will be a minimum variance unbiased estimator of p.

The conditional distribution of X_1, given $\bar{X}$ and S, is derived from the fact that

$$\frac{X_1 - \bar{X}}{S} \left(\frac{n}{n-1} \right)^{\frac{1}{2}}$$

is distributed as t_{n-1} (t with $(n-1)$ degrees of freedom) (section 1.7). Hence the conditional distribution of X_1, given $\bar{X}$ and S, is that of

$$\bar{X} + \left(\frac{n-1}{n} \right)^{\frac{1}{2}} S t_{n-1}$$

(In this expression, $\bar{X}$ and S are to be regarded as constants.) Hence

$$\tilde{p} = \text{E}[\hat{p} | \bar{X}, S] = 1 - \Pr\left[\frac{\text{LSL} - \bar{X}}{S} \left(\frac{n}{n-1} \right)^{\frac{1}{2}} \right.$$

$$\left. < t_{n-1} < \frac{\text{USL} - \bar{X}}{S} \left(\frac{n}{n-1} \right)^{\frac{1}{2}} \right]$$

$$= 1 - I_{\theta_1(\bar{X}, S)}(\tfrac{1}{2}(n-1), \tfrac{1}{2}(n-1))$$

$$+ I_{\theta_2(\bar{X}, S)}(\tfrac{1}{2}(n-1), \tfrac{1}{2}(n-1)) \tag{2.52}$$

where $I_\theta(a, b)$ denotes incomplete beta function ratio (section 1.6) and

$$\theta_1(\bar{X}, S) = \frac{1}{2}\left[1 + \frac{USL - \bar{X}}{S}\left\{\frac{(n-1)^2}{n} + \left(\frac{USL - \bar{X}}{S}\right)^2\right\}^{-\frac{1}{2}}\right]$$

$$\theta_2(\bar{X}, S) = \frac{1}{2}\left[1 + \frac{LSL - \bar{X}}{S}\left\{\frac{(n-1)^2}{n} + \left(\frac{LSL - \bar{X}}{S}\right)^2\right\}^{-\frac{1}{2}}\right]$$

Calculation of $\tilde{p}$ is quite heavy, but can be facilitated by tables of the function

$$J(n, c) = I_{\frac{1}{2}[1 + c\{\frac{(n-1)^2}{n} + c^2\}^{-\frac{1}{2}}]}(\tfrac{1}{2}(n-1), \tfrac{1}{2}(n-1)) \qquad (2.53\ a)$$

for various values of n and c. Then

$$\hat{p} = 1 - J\left(n, \frac{USL - \bar{X}}{S}\right) + J\left(n, \frac{LSL - \bar{X}}{S}\right) \qquad (2.53\ b)$$

It must be emphasized that the appropriateness of both of the estimators $\tilde{p}$ and $\hat{p}$ depends on the correctness of the assumption of normality for the process distribution.

APPENDIX 2.B

$X_1, X_2, \ldots, X_n$ are dependent random variables with common expected value ξ and variance σ^2; the correlation between X_i and X_j will be denoted by ρ_{ij}. We have

$$\text{var}(\bar{X}) = n^{-2}\sigma^2\left\{n + n(n-1)\sum_{i \neq j} \rho_{ij}\right\} = \frac{1}{n}\{1 + (n-1)\bar{\rho}\}\sigma^2$$

where

$$\bar{\rho} = \frac{\displaystyle\sum_{i \neq j}\sum \rho_{ij}}{n(n-1)}$$

If

$$S^2 = (n-1)^{-1} \sum_{j=1}^{n} (X_j - \bar{X})^2$$

then

$$E[S^2] = \frac{1}{n-1} \sum_{j=1}^{n} E[(X_j - \bar{X})^2]$$

$$= \frac{1}{n-1} \sum_{j=1}^{n} E[(X_j - \xi)^2] - 2(\bar{X} - \xi)(X_j - \xi) + (\bar{X} - \xi)^2]$$

$$= \frac{1}{n-1} \left[\sum_{j=1}^{n} E[(X_j - \xi)^2] - nE(\bar{X} - \xi)^2 \right]$$

since $\Sigma(X_j - \xi) = n(\bar{X} - \xi)$ so

$$E[S^2] = \frac{1}{n-1} \{n \, \mathrm{var}(X_j) - n \, \mathrm{var}(\bar{X})\}$$

$$= \frac{1}{n-1} [n - \{1 + (n-1)\bar{\rho}\}]\sigma^2$$

$$= (1 - \bar{\rho})\sigma^2 \qquad\qquad (2.54)$$

We emphasize that

$$\bar{\rho} = \frac{1}{n(n-1)} \sum_{i \neq j}\sum \rho_{ij}$$

is the average of the correlation coefficients among the $\frac{1}{2}n(n-1)$ different pairs of Xs.

BIBLIOGRAPHY

Barnett, N.S. (1988) *Process control and product quality: The C_p and C_{pk} revisited.* Footscray Inst. Technol., Victoria, Australia.

Bian, G. and Saw, S.L.C. (1993) Bayes estimates for two functions of the normal mean and variance with a quality control application, *Tech. Rep.*, National University of Singapore.

Bissell, A.F. (1990) How reliable is your capability index? *Appl. Statist.*, **39**, 331–40.

Bunks, J. (1989) *Principles of Quality Control*, Wiley: New York.

Carr, W.E. (1991) A new process capability index: parts per million, *Quality Progress*, **24**, (2), 152.

Chen, L.K., Xiong, Z. and Zhang, D. (1990) On asymptotic distributions of some process capability indices, *Commun. Statist. – Theor. Meth.*, **19**, 11–18.

Chou, Y.M. and Owen, D.B. (1989) On the distribution of the estimated process capability indices, *Commun. Statist. – Theor. Meth.*, **18**, 4549–60.

Chou, Y.M. and Owen, D.B. (1991) A likelihood ratio test for the equality of proportions of two normal populations, *Commun. Statist. – Theor. Meth.*, **20**, 2357–2374.

Chou, Y.M., Owen, D.B. and Borrego, A.S.A. (1990) Lower confidence limits on process capability indices, *J. Qual. Technol*, **22**, 223–9.

Clements, R.R. (1988) *Statistical Process Control*, Kriger: Malabar, Florida.

Coleman, D.E. (1991) Relationships between Loss and Capability Indices, *Apl. Math. Comp. Techn.*, ALCOA Techn. Center, PA.

Constable, G.K. and Hobbs, J.R. (1992) Small Samples and Non-normal Capability, *Trans. ASQC Quality Congress*, 1–7.

Dovich, R.A. (1992) Private communications.

Elandt-Johnson, R.C. and Johnson, N.L. (1980) *Survival Models and Data Analysis*, Wiley: New York.

Fleiss, J.L. (1973) *Statistical Methods for Rates and Proportions*, Wiley: New York.

Franklin, L.A. and Wasserman, G.S. (1992). A note on the conservative nature of the tables of lower confidence limits for C_{pk} with a suggested correction, *Commun. Statist. Simul. Comp.*

Gensidy, A. (1985) C_p and C_{pk}, *Qual. Progress*, 18(4), 7–8.

Grant, E.L. and Leavenworth, R.S. (1988) *Statistical Quality Control* (6th edn.), McGraw-Hill, New York.

Guirguis, G. and Rodriguez, R.N. (1992) Computation of Owen's Q function applied to process capability analysis, *J. Qual. Technol.*, **24**, 236–246.

Heavlin, W.D. (1988) Statistical properties of capability indices, *Technical Report No. 320*, Tech. Library, Advanced Micro Devices, Inc., Sunnyvale, California.

Herman, J.T. (1989) Capability index – enough for process industries? *Trans. ASQC Congress, Toronto*, 670–5.

John, P.W.M. (1990) *Statistical Methods in Engineering and Quality Assurance*, Wiley: New York.

Johnson, M. (1992) Statistics simplified. *Qual. Progress*, **25**(1), 10–11.

Kane, V.E. (1986) Process capability indices, *J. Qual. Technol.*, **18**, 41–52.

Kirmani, S.N.U.A., Kocherlakota, K. and Kocherlakota, S. (1991) Estimation of $\bar{\sigma}$ and the process capability index based on subsamples, *Commun. Statist. – Theor. Meth.*, **20**, 275–291 (Correction, **20**, 4083).

Kotz, S., Pearn, W.L. and Johnson, N.L. (1992) Some process capability indices are more reliable than one might think, *Appl. Statist.*, **42**, 55–62.

Kushler, R. and Hurley, P. (1992) Confidence bounds for capability indices, *J. Qual. Technol.*, **24**, 188–195.

Lam, C.T. and Littig, S.J. (1992) A new standard in process capability measurement. *Tech. Rep. 92–93*, Dept. Industrial and Operations Engineering, University of Michigan, Ann Arbor.

Leone, F.C., Nelson, L.S., and Nottingham, R.R. (1961) The folded normal distribution, *Technometrics*, **3**, 543–50.

Li, H., Owen, D.B. and Borrego, A.S.A. (1990) Lower confidence limits on process capability indices based on the range, *Commun. Statist. – Simul. Comp.*, **19**, 1–24.

Marcucci, M.O. and Beazley, CC. (1988) Capability indices: Process performance measures, *Trans. ASQC Congress*, 516–23.

Montgomery, D.C. (1985) *Introduction to Statistical Quality Control*, Wiley: New York.

Nagata, Y. (1991) Interval estimation for the process capability indices, *J. Japan. Soc. Qual. Control*, **21**, 109–114.

Nagata, Y. and Nagahata, H. (1993) Approximation formulas for the confidence intervals of process capability indices, *Tech. Rep.* Okayama University, Japan, (submitted for publication).

Owen, D.B. (1965) A special case of a bivariate noncentral *t*-distribution, *Biometrika*, **53**, 437–46.

Owen, M. (ed.) (1989) *SPC and Continuous Improvement*, Springer Verlag, Berlin, Germany.

Pearn, W.L., Kotz, S. and Johnson, N.L. (1992) Distributional and inferential properties of process capability indices, *J. Qual. Technol.*, **24**, 216–231.

Porter, L.J. and Oakland, J.S. (1990) Measuring process capability using indices – some new considerations, *Qual. Rel. Eng. Internat.*, **6**, 19–27.

Wadsworth, H.M., Stephens, K.S. and Godfrey, A.B. (1988) *Modern Methods for Quality Control and Improvement*, Wiley: New York.

Wierda, S.J. (1992) A multivariate process capability index, *Proc. 9th Internat. Conf. Israel Soc. Qual. Assur.* (Y. Bester, H. Horowitz and A. Lewis, eds.) pp. 517–522.

Yang, K. and Hancock, W.M. (1990) Statistical quality control for correlated samples, *Intern. J. Product. Res.*, **28**, 595–608.

Zhang, N.F., Stenback, G.A. and Wardrop, D.M. (1990) Interval estimation of process capability index C_{pk}. *Commun. Statist. – Theor. Meth.*, 19, 4455–4470.

3

The C_{pm} index and related indices

3.1 INTRODUCTION

The C_{pm} index of process capability was introduced in the published literature by Chan *et al.* (1988 a). An elementary discussion of the index, written for practitioners, was presented by Spiring (1991 a), who also (Spiring, 1991 b) extended the application of the index to measure process capability in the presence of 'assignable causes'. He had previously (Spiring, 1989) described the application of C_{pm} to a toolwear problem, as an example of a process in which there was variation due to assignable causes.

Recent papers (unpublished at the time of writing) in which the use of C_{pm} is discussed include Subbaiah and Taam (1991) and Kushler and Hurley (1992). (All four of these authors were in the Department of Mathematical Sciences, Oakland University, Rochester, Minnesota.) These papers study estimation of C_{pm} in some detail; graphical techniques for estimating C_{pm} are discussed briefly by Chan *et al.* (1988 b).

The use of the C_{pm} index was proposed by Hsiang and Taguchi (1985) at the American Statistical Association's Annual Meeting in Las Vegas, Nevada (though they did not use the symbol C_{pm}). Their approach was motivated exclusively by loss function considerations, but Chan *et al.* (1988 a, b) were more concerned with comparisons with C_{pk}, discussed in Chapter 2. Marcucci and Beazley (1988), apparently independently, but quite possibly influenced by Hsiang and Taguchi, proposed the equivalent index, also defined explicitly by Chan *et al.* (1988 a)

$$C_{pg} = C_{pm}^{-2}$$

(proportional to average quadratic loss). Johnson (1991) – see section 3.2 – also reaches 'L_e', a multiple of C_{pm}^{-2} from loss – or, conversely, 'worth' – function considerations.

Chan *et al.* (1988 b) define a slightly different index, C_{pm}^* (see (3.2) below), which is more closely related, in spirit and form, to C_{pk} than is C_{pm}.

Finally an index C_{pmk}, combining features of C_{pm} and C_{pk} is introduced, and studied in section 3.5.

3.2 THE C_{pm} INDEX

The original definition of C_{pm} is

$$C_{pm} = \frac{\text{USL} - \text{LSL}}{6\{\sigma^2 + (\xi - T)^2\}^{\frac{1}{2}}} = \frac{d}{3\{\sigma^2 + (\xi - T)^2\}^{\frac{1}{2}}} \qquad (3.1)$$

where ξ is the expected value and σ is the standard deviation of the measured characteristic, X, T is the target value and $d = \frac{1}{2}(\text{USL} - \text{LSL})$. The target value is commonly the midpoint of the specification interval, $m = \frac{1}{2}(\text{USL} + \text{LSL})$. This need not be the case, but if it is not so, there can be serious pitfalls in the uncritical use of C_{pm}, as we shall see later.

The modified index, C_{pm}^*, is defined using (1.14 e) by

$$C_{pm}^* = \frac{\min(\text{USL} - T, T - \text{LSL})}{3\{\sigma^2 + (\xi - T)^2\}^{\frac{1}{2}}} = \frac{d - |T - m|}{3\{\sigma^2 + (\xi - T)^2\}^{\frac{1}{2}}} \qquad (3.2)$$

If $T=m$ (as is often the case), then $C^*_{pm}=C_{pm}$. If $T\neq m$, the same type of pitfalls beset uncritical use of C^*_{pm} as with C_{pm}.

The denominator in each of (3.1) and (3.2)

$$3\{\sigma^2+(\xi-T)^2\}^{\frac{1}{2}}=3\{E[(X-T)^2]\}^{\frac{1}{2}} \qquad (3.3)$$

that is, three times the root-mean-square deviation of X from the target value, T. It is clear that, since

$$\sigma^2+(\xi-T)^2\geqslant\sigma^2$$

we must have

$$C_{pm}\leqslant C_p \qquad (3.4)$$

The relationship between C_{pm} and C_{pk} is discussed below; see (3.10) *et seq.*

In the approach of Hsiang and Taguchi (1985) it is supposed that the loss function is proportional to $(X-T)^2$, and so $\sigma^2+(\xi-T)^2$ is a measure of average loss. From this point of view, there is no need to include USL or LSL in the numerator, or, indeed, the multiplier 6 in the denominator, of the formula for C_{pm}, except to maintain comparability with the earlier index C_p, which was introduced with quite a different background, although for the same purpose.

Proponents of Hsiang and Taguchi's approach point out that it does allow for differing costs arising from different values of X in the range LSL to USL, though in a necessarily arbitrary way. (A discussion between Mirabella (1991) and Spiring (1991 b) is relevant to this point.) It is suggested that a reasonable loss function might be constant outside this range, perhaps giving a function of form

$$\begin{cases} c(X-T)^2 & \text{for } LSL\leqslant X\leqslant USL \\ c(LSL-T)^2 & \text{for } X\leqslant LSL \\ c(USL-T)^2 & \text{for } X\geqslant USL \end{cases} \qquad (3.5)$$

One is, of course, free to use one's intuition *ad libitum* in construction of a loss function. However, if this approach is accepted, there seems to be little justification, in respect of practical value, ease of calculation, or simplicity of mathematical theory in using the cumbersome C_{pm} (or C_{pm}^*) form rather than the loss function itself (as in, for example, C_{pg} or L_e – see sections 3.1 and 3.4).

As already remarked, we take the position that expected proportion of NC product is of primary importance, though we acknowledge that other factors can be of importance in specific cases. Essentially, the use of any single index cannot be entirely satisfactory, though it can be of value in rapid, informal communication or preliminary, cursory assessment of a situation.

The following interrelations should be noted. We have

$$C_{pm} = \frac{1}{3} \frac{d}{\sigma'} \tag{3.6}$$

where $\sigma' = \{E[(X-T)^2]\}^{\frac{1}{2}}$ measures the 'total variability' about the target value. Recalling the definition (in Chapter 2)

$$C_p = \frac{\text{USL} - \text{LSL}}{6\sigma} = \frac{1}{3} \frac{d}{\sigma} \tag{3.7}$$

we have

$$C_{pm} = \frac{\sigma}{\sigma'} C_p = C_p (1 + \zeta^2)^{-\frac{1}{2}} \tag{3.8}$$

where $\zeta = (\xi - T)/\sigma$. So $C_{pm} \leqslant C_p$ with equality if $\xi = T$.

Boyles (1991) provides an instructive comparison between C_p and C_{pm}. He takes USL $= 65$, LSL $= 35$, $T = 50$

$(=\frac{1}{2}(\text{USL}+\text{LSL})=m)$ so that $d=\frac{1}{2}(65-35)=15$, and considers three processes with the following characteristics:

A: $\xi=50(=T)$; $\sigma=5$
B: $\xi=57.5$; $\sigma=2.5$
C: $\xi=61.25$; $\sigma=1.25$

The values of the indices C_p, C_{pk} and C_{pm} are

Process	C_p	C_{pk}	C_{pm}
A	1	1	1
B	2	1	0.63
C	4	1	0.44

$(C_{pm}^*=C_{pm}$, since $T=m$).

Note that, in this example, C_p and C_{pm} move in opposite directions. According to C_p, the processes are, in order of desirability – C:B:A – while for C_{pm} the order is reversed. According to the index C_{pk}, there is nothing to choose among the processes.

At this point we repeat our warning that C_{pm}, as defined in (3.1), is not a satisfactory measure of process capability unless the target value is equal to the midpoint of the specification interval (i.e. $T=m=\frac{1}{2}(\text{USL}+\text{LSL})$).

The reason for this is that, whether the expected value of X is $T-\delta$ or $T+\delta$, we obtain the same value of C_{pm}, namely

$$\frac{1}{3}d(\sigma^2+\delta^2)^{-\frac{1}{2}},\qquad(3.9)$$

although, unless $T=m$, these processes can have markedly different expected proportions of NC items (items with X outside specification limits). For example, if

$$T=\tfrac{3}{4}\text{USL}+\tfrac{1}{4}\text{LSL}$$

and

$$\delta = \tfrac{1}{4}(\mathrm{USL} - \mathrm{LSL}) = \tfrac{1}{2}d$$

then with $\mathrm{E}[X] = T + \delta = \mathrm{USL}$, we would expect at least 50% NC items if X has a symmetrical distribution, while when $\mathrm{E}[X] = T - \delta = m$, this proportion would, in general, be much less.

In the light of this, it is, in our opinion, pointless to discuss properties of C_{pm} (as defined in (3.1)) except when $T = m$. These comments also apply to C_{pm}^{*}.

Generally, it is not possible to calculate C_{pm} from C_{pk}, or conversely, without knowledge of values of the ratios $d : \sigma : |\xi - m|$.

If $T = m$ then from (3.1), and (2.20 b)

$$\frac{C_{\mathrm{pk}}}{C_{\mathrm{pm}}} = \left\{ 1 - \frac{|\xi - m|}{d} \right\} \times \left[1 + \left\{ \frac{\xi - m}{\sigma} \right\}^{2} \right]^{\frac{1}{2}} \qquad (3.10)$$

If, further $C_{\mathrm{p}} = 1$ (i.e. $d = 3\sigma$) then

$$9(1 - C_{\mathrm{pk}})^{2} = \left(\frac{\xi - m}{\sigma} \right)^{2} = C_{\mathrm{pm}}^{-2} - 1 \qquad (3.11)$$

Since (c.f. (2.20 c))

$$C_{\mathrm{pk}} = \left(1 - \frac{|\xi - m|}{d} \right) C_{\mathrm{p}} \qquad C_{\mathrm{pm}} = \left\{ 1 + \left(\frac{\xi - T}{\sigma} \right)^{2} \right\}^{-\frac{1}{2}} C_{\mathrm{p}}$$

we have

$$C_{\mathrm{p}} \geqslant \max\,(C_{\mathrm{pk}}, C_{\mathrm{pm}})$$

Also, $C_{\mathrm{pk}} \geqslant$ or $\leqslant C_{\mathrm{pm}}$ according as

$$1 - \frac{|\xi - m|}{d} \geqslant \quad \text{or} \quad \leqslant \left\{ 1 + \left(\frac{\xi - T}{\sigma} \right)^{2} \right\}^{-\frac{1}{2}}$$

i.e. as

$$\left(1-\frac{|\xi-m|}{d}\right)\left\{1+\left(\frac{\xi-T}{\sigma}\right)^2\right\}^{\frac{1}{2}} \geqslant \quad \text{or} \quad \leqslant 1 \qquad (3.12)$$

If $m = T$, then (3.12) becomes

$$(1-u)\left\{1+\left(\frac{d}{\sigma}\right)^2 u^2\right\}^{\frac{1}{2}} \geqslant \quad \text{or} \quad \leqslant 1$$

(where

$$u = \frac{|\xi-m|}{d} = \frac{|\xi-T|}{d}\Big);$$

i.e.

$$1+\left(\frac{d}{\sigma}\right)^2 u^2 \geqslant \quad \text{or} \quad \leqslant (1-u)^{-2}$$

$$\left(\frac{d}{\sigma}\right)^2 \geqslant \quad \text{or} \quad \leqslant \frac{(1-u)^{-2}-1}{u^2} = \frac{1-(1-u)^2}{u^2(1-u)^2} = \frac{2-u}{u(1-u)^2}$$

$$(3.13)$$

Usually $u < 1$ (or else ξ is outside specification limits).
Inequalities (3.13) can be written

$$C_{\text{p}}^2 \geqslant \quad \text{or} \quad \leqslant \frac{2-u}{9u(1-u)^2}$$

Here are some values of this function.

u	0	0.2	0.4	0.5	0.6	0.8	1
$\dfrac{2-u}{9u(1-u)^2}$	∞	1.56	1.23	1.33	1.62	4.17	∞

This function has a minimum value of 1.23 at $u=0.38$ (approx.). If C_p is unsatisfactory ($C_p \leqslant 1$) then we always have $C_{pk} \leqslant C_{pm}$ (if $T=m$), but we can have $C_{pk} > C_{pm}$ when $C_p^2 > 1.23 (C_p > 1.11)$, provided u is not too close to 0 or 1. If $u=0$ (i.e. $\xi = T = m$) then $C_p = C_{pk} = C_{pm}$.

Johnson (1991) also approached the C_{pm} index from the point of view of loss functions, using the converse concept of 'worth'. He supposed that the maximum worth, W_T, of an item corresponds to the characteristic X having the target value, T. As the deviation of X from T increases, the worth becomes less, eventually becoming zero, and then negative. If the worth function is of the simple form

$$W_T - k(X-T)^2 \quad \text{for } W_T \geqslant k(X-T)^2$$

it becomes zero when $|X-T| = \sqrt{W_T/k}$. In Johnson's (1991) approach the values $T \pm \sqrt{W_T/k}$ play similar roles to those of the specification limits in defining C_{pm} and we may define $\Delta = \sqrt{W_T/k}$.

The ratio (worth)/(maximum worth)

$$= 1 - \frac{k(X-T)^2}{W_T}$$

$$= 1 - \frac{(X-T)^2}{\Delta^2}$$

is termed rhe **relative worth**, and $(X-T)^2/\Delta^2$ is termed the **relative loss**. The expected relative loss is then (in Johnson's notation)

$$\frac{E[(X-T)^2]}{\Delta^2} = L_e$$

Recalling that

$$C^2_{pm} = \left(\frac{d}{3}\right)^2 \{E[X-T)^2]\}^{-1}$$

we see that,

$$L_e = \left(\frac{d}{3\Delta}\right)^2 C^{-2}_{pm}$$

Hence L_e is effectively equivalent to C_{pm}, although it has a different background. A natural unbiased estimator of L_e is

$$\hat{L}_e = \frac{1}{n\Delta^2} \sum_{i=1}^{n} (X_i - T)^2$$

The distribution of $n\Delta^2 \hat{L}_e$ is therefore that of

$$\sigma^2 \chi'^2_n \left(\frac{n(\xi-T)^2}{\sigma^2}\right)$$

discussed in section 1.5. The resulting analysis for construction of confidence intervals follows lines to be developed in section 3.5. Extension to nonsymmetric loss functions is direct.

3.3 COMPARISON OF C_{pm} AND C^*_{pm} WITH C_p AND C_{pk}

We first note that if $T = \xi$, then C_{pm} is the same as C_p. If, also, $\xi = m$, then both C_{pm} and C^*_{pm} are the same as C_{pk}. If, further, the process characteristic has a 'stable normal distribution' then a value of 1 (for all four indices) means that the expected proportion of product with values of X within specification limits is a little over 99.7%. Negative values are impossible for C_p and C_{pm}. They are impossible for C_{pm}^* if T falls within specification limits (and for C_{pk} if ξ falls within specification limits).

The principal distinction between C_{pm} and C_{pk} is, as Mirabella (1991) implicitly indicates, and Kushler and Hurley (1992) state explicitly, in the relative importance of the specification limits (USL and LSL) versus the target value, T. As explained in Chapter 2, the main function of C_{pk} is to indicate the degree to which the process is within specification limits. For $LSL < \xi < USL$, $C_{pk} \to \infty$ as $\sigma \to 0$, but large values of C_{pk} do not provide information about discrepancy between ξ and T. The index C_{pm}, however, purports to measure the degree of 'on-targetness' of the process, and $C_{pm} \to \infty$ if $\xi \to T$, as well as $\sigma \to 0$. With C_{pm}, specification limits are used solely in order to scale the loss function (squared deviation) in the denominator.

Kushler and Hurley (1992, p. 2) emphasize that, if T is not equal to m, the following paradox can arise '...moving the process mean towards the target (which will increase the C_{pk}† and C_{pm} indices) can reduce the fraction of the distribution which is within specification limits.'

Finally, we compare plots of values of (ξ, σ) corresponding to fixed values of C_p, C_{pk} and C_{pm}.

If $C_{pm} = c$, then

$$\sigma^2 + (\xi - T)^2 = \left(\frac{1}{3} \frac{d}{c} \right)^2$$

These (ξ, σ) points lie on a semi-circle with centre at $(T, 0)$, radius $\frac{1}{3} d/c$, and restricted to non-negative σ.

---

†The C_{pk} index referred to here is the modified index described by Kane (1986), namely $\frac{1}{3}\{d - |\xi - T|\}/\sigma$.

If $C_p = c$, we have (ξ, σ) points on the line

$$\sigma = \frac{1}{3}\frac{d}{c}$$

parallel to the ξ-axis, and tangential to the semi-circle corresponding to $C_{pm} = c$ at the point $(T, \frac{1}{3}d/c)$.

If $C_{pk} = c$ we have

$$|\xi - m| + 3c\sigma = d$$

(and $\sigma \geqslant 0$). This corresponds to (ξ, σ) points on the pair of lines

$$\xi = m + d - 3c\sigma$$

and

$$\xi = m - d + 3c\sigma$$

These lines intersect at the point $(m, \frac{1}{3}d/c)$ and meet the ξ-axis at the points $(m-d, 0)$ and $(m+d, 0)$. Boyles (1991, Figs. 4 and 5) provides diagrams representing these loci from the example already described in section 3.2 (USL = 65, LSL = 35, $m = 50$, $2d = 30$) with $c = \frac{1}{3}, \frac{2}{3}, 1$ and $\frac{4}{3}$.

Figure 3.1 is a simplified version, giving (ξ, σ) loci for a general value of c, exceeding $\frac{1}{3}$ with $T = m$. (If $c < \frac{1}{3}$ the lines $C_{pk} = c$ would meet the ξ-axis between its points of intersection $(\xi = m \pm \frac{1}{3}d/c)$ with the $C_{pm} = c$ semi-circle.) If $T \neq m$, the lines $C_{pk} = c$ and $C_p = c$ would be unchanged but the $C_{pm} = c$ semi-circle would be moved horizontally to centre at the point $(T, 0)$, with points of intersection with the ξ-axis at $\xi = T \pm \frac{1}{3}d/c$.

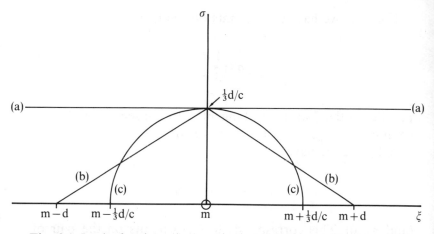

Fig. 3.1 Loci of points (ξ, σ) such that: (a) $C_p = c$; (b) $C_{pk} = c$; (c) $C_{pm} = c$; when $T = m$, $c > \frac{1}{3}$. (If $c < \frac{1}{3}$, $m - d/3c < m - d < m + d < m + d/3c$.) If $c = \frac{1}{3}$ then $\sigma = d$.

3.4 ESTIMATION OF C_{pm} (AND C_{pm}^*)

3.4.1 Point estimation

A natural estimator of C_{pm} is obtained by replacing $E[(X - T)^2]$ by its unbiased estimator

$$\tilde{\sigma}'^2 = \frac{1}{n} \sum_{i=1}^{n} (X_i - T)^2 \qquad (3.14)$$

Of course, $\tilde{\sigma}'$ is not an unbiased estimator of σ', but has a negative bias, since $E[\tilde{\sigma}'^2] > \{E[\tilde{\sigma}']\}^2$. Nor is $\frac{1}{3}d/\tilde{\sigma}'$ an unbiased estimator of C_{pm}.

We define

$$\tilde{C}_{pm} = \frac{d}{3\tilde{\sigma}'} = \frac{d}{3} \left\{ \frac{1}{n} \sum_{i=1}^{n} (X_i - T)^2 \right\}^{-\frac{1}{2}} \qquad (3.15)$$

The corresponding estimator of C_{pm}^* is

$$\tilde{C}_{\mathrm{pm}}^* = \frac{(d-|T-m|)}{3} \left\{ \frac{1}{n} \sum_{i=1}^n (X_i - T)^2 \right\}^{-\frac{1}{2}} \tag{3.16}$$

This, also, is a biased estimator. The biases of both $\tilde{C}_{\mathrm{pm}}$ and $\tilde{C}_{\mathrm{pm}}^*$ will generally be positive (because the bias of $\tilde{\sigma}'$ is negative).

For the case when each of $X_1, \ldots, X_n$ has a normal distribution with expected value ξ and standard deviation σ, the exact distribution of $\tilde{C}_{\mathrm{pm}}$ (and of $\tilde{C}_{\mathrm{pm}}^*$) can be derived. The distribution of

$$n\tilde{\sigma}'^2 = \sum_{i=1}^n (X_i - T)^2$$

is that of $\sigma^2 \times$ (noncentral χ^2 with n degrees of freedom and noncentrality parameter $\lambda = n(\xi - T)^2 \sigma^{-2}$), symbolically

$$\sigma^2 \chi_n'^2(\lambda) \tag{3.17}$$

This is a mixture of (central) chi-squared distributions with $(n+2j)$ degrees of freedom and corresponding Poisson weights

$$P_j(\tfrac{1}{2}\lambda) = \exp(-\tfrac{1}{2}\lambda) \frac{(\tfrac{1}{2}\lambda)^j}{j!} \quad j = 0, 1, \ldots \tag{3.18}$$

(See section 1.5.2).

The expected value of $\tilde{C}_{\mathrm{pm}}^r$ (the rth moment of $\tilde{C}_{\mathrm{pm}}$ about zero) is

$$\mu_r'(\tilde{C}_{\mathrm{pm}}) = \left(\frac{d\sqrt{n}}{3\sigma}\right)^r \exp(-\tfrac{1}{2}\lambda) \sum_{j=0}^\infty \frac{(\tfrac{1}{2}\lambda)^j}{j!} E[(\chi_{n+2j}^2)^{-\frac{1}{2}r}]$$

$$= \left(\frac{d\sqrt{n}}{3\sigma}\right)^r \exp(-\tfrac{1}{2}\lambda) \sum_{j=0}^\infty \frac{(\tfrac{1}{2}\lambda)^j}{j!} \frac{\Gamma(\tfrac{1}{2}(n-r)+j)}{2^{\frac{1}{2}r}\Gamma(\tfrac{1}{2}n+j)} \tag{3.19}$$

(For $r \geqslant n$, $\mu_r'(\tilde{C}_{\mathrm{pm}})$ is infinite.)

The rth moment about zero of $\tilde{C}^*_{pm}$ is obtained from (3.19) by replacing d by $d - |T - m|$.

Taking $r = 1$ and $r = 2$ we obtain

$$\mu'_1(\tilde{C}_{pm}) = E[\tilde{C}_{pm}] = \frac{d\sqrt{n}}{3\sigma} \exp(-\tfrac{1}{2}\lambda) \sum_{j=0}^{\infty} \frac{(\tfrac{1}{2}\lambda)^j}{j!} \frac{\Gamma(\tfrac{1}{2}(n-1)+j)}{2^{\frac{1}{2}}\Gamma(\tfrac{1}{2}n+j)}$$

$$(3.20\,a)$$

and

$$\text{var}(\tilde{C}_{pm}) =$$

$$\left(\frac{d\sqrt{n}}{3\sigma}\right)^2 \exp(-\tfrac{1}{2}\lambda) \sum_{j=0}^{\infty} \frac{(\tfrac{1}{2}\lambda)^j}{j!} \frac{1}{n-2+2j}$$

$$- \{E[\tilde{C}_{pm}]\}^2$$

$$(3.20\,b)$$

Table 3.1 gives values of $E[\tilde{C}_{pm}]$ and S.D. $(\tilde{C}_{pm}) = (\text{var}(\tilde{C}_{pm}))^{\frac{1}{2}}$ for selected values of n, d/σ and $|\xi - T|/\sigma (=\sqrt{\lambda/n})$. (See also the note on calculation of these quantities, at the end of this section.)

Table 3.2 gives corresponding values of C_{pm}.

When $\xi = T$ we have $\lambda = 0$ and

$$C_{pm} = \frac{d}{3\sigma} \; (= C_p)$$

$$(3.21\,a)$$

$$E[\tilde{C}_{pm}] = \frac{d\sqrt{n}}{3\sigma} \frac{\Gamma(\tfrac{1}{2}(n-1))}{2^{\frac{1}{2}}\Gamma(\tfrac{1}{2}n)}$$

$$(3.21\,b)$$

$$\text{var}(\tilde{C}_{pm}) = \left(\frac{d\sqrt{n}}{3\sigma}\right)^2 \left[\frac{1}{n-2} - \frac{1}{2}\left\{\frac{\Gamma(\tfrac{1}{2}(n-1))}{\Gamma(\tfrac{1}{2}n)}\right\}^2\right]$$

$$(3.21\,c)$$

cf. (2.16 a) with $f = n$, and the bias of $\tilde{C}_{pm}$, as an estimator of C_{pm}, is

$$E[\tilde{C}_{pm}] - C_{pm} = \frac{d}{3\sigma} \left[\left(\frac{n}{2}\right)^{\frac{1}{2}} \frac{\Gamma(\tfrac{1}{2}(n-1))}{\Gamma(\tfrac{1}{2}n)} - 1\right]$$

$$(3.22)$$

Since $\Gamma(y-\frac{1}{2})\sqrt{y}/\Gamma(y) > 1$ for all $y \geqslant 1$ the bias is positive, as expected – see the remarks following (3.16).

Some authors, for example, Spiring (1989) use the estimator

$$\hat{C}_{pm} = \frac{d}{3\left\{\dfrac{1}{n-1} \displaystyle\sum_{i=1}^{n} (X_i - T)^2\right\}^{\frac{1}{2}}} = \left(\frac{n-1}{n}\right)^{\frac{1}{2}} \tilde{C}_{pm} \quad (3.23\,a)$$

in place of $\tilde{C}_{pm}$, (replacing the divisor n by $(n-1)$, in the denominator). This is presumably inspired by analogy with the unbiased estimator of σ^2,

$$S^2 = \frac{1}{n-1} \sum_{i=1}^{n} (X_i - \bar{X})^2$$

Since $\hat{C}_{pm} < \tilde{C}_{pm}$, the bias of $\hat{C}_{pm}$ must be less than that of $\tilde{C}_{pm}$. Although the bias of $\tilde{C}_{pm}$ is positive, however, that of $\hat{C}_{pm}$ can be negative, though it cannot be less than

$$-C_{pm}\left(1 - \left(\frac{n-1}{n}\right)^{\frac{1}{2}}\right) \qquad (3.23\,b)$$

(Subbaiah and Taam, 1991). (This follows from the fact that

$$E[\hat{C}_{pm}] - C_{pm} = \left(\frac{n-1}{n}\right)^{\frac{1}{2}} E[\tilde{C}_{pm}] - C_{pm}$$

$$> \left(\frac{n-1}{n}\right)^{\frac{1}{2}} C_{pm} - C_{pm}$$

because $E[\hat{C}_{pm}] > C_{pm}$.)

If the bias of $\hat{C}_{pm}$ (as well as that of $\tilde{C}_{pm}$) is positive, and also $\xi = T$, then the mean square error of $\hat{C}_{pm}$ is less than that of $\tilde{C}_{pm}$, because both the variance and the squared bias of $\hat{C}_{pm}$ are less than those of $\tilde{C}_{pm}$. It should be noted, however, that this

Table 3.1 Moments of C_{pm}: E = E[$\tilde{C}_{pm}$]; S.D. = S.D. ($\tilde{C}_{pm}$)

					$|\xi - T|/\sigma$					
	0		0.5		1		1.5		2	
	E	S.D.	E	S.D.	E	S.D.	E	S.D.	E	S.D.
$n=10$										
d/σ										
2	0.722	0.183	0.644	0.161	0.501	0.110	0.386	0.069	0.307	0.044
3	1.084	0.275	0.967	0.241	0.752	0.165	0.578	0.103	0.460	0.066
4	1.445	0.366	1.289	0.322	1.002	0.220	0.771	0.137	0.613	0.088
5	1.806	0.458	1.611	0.402	1.253	0.275	0.964	0.171	0.767	0.110
6	2.167	0.550	1.933	0.482	1.503	0.330	1.157	0.206	0.920	0.132
$n=15$										
d/σ										
2	0.702	0.139	0.627	0.122	0.490	0.084	0.380	0.053	0.304	0.035
3	1.054	0.209	0.941	0.183	0.736	0.126	0.570	0.080	0.456	0.052
4	1.405	0.278	1.254	0.244	0.981	0.168	0.760	0.106	0.607	0.070
5	1.756	0.348	1.568	0.305	1.226	0.210	0.950	0.133	0.759	0.087
6	2.107	0.417	1.881	0.366	1.471	0.252	1.140	0.160	0.911	0.104

$n=20$										
d/σ										
2	0.693	0.116	0.619	0.102	0.485	0.070	0.377	0.045	0.302	0.030
3	1.040	0.174	0.928	0.153	0.728	0.106	0.566	0.068	0.453	0.044
4	1.386	0.233	1.238	0.204	0.971	0.141	0.755	0.090	0.605	0.059
5	1.733	0.291	1.547	0.255	1.214	0.176	0.943	0.113	0.756	0.074
6	2.079	0.349	1.857	0.306	1.456	0.211	1.132	0.135	0.907	0.089

$n=25$										
d/σ										
2	0.688	0.102	0.614	0.089	0.482	0.062	0.376	0.040	0.301	0.026
3	1.031	0.153	0.921	0.134	0.724	0.093	0.564	0.060	0.452	0.039
4	1.375	0.204	1.228	0.179	0.965	0.124	0.752	0.079	0.603	0.052
5	1.719	0.255	1.536	0.223	1.206	0.155	0.939	0.099	0.754	0.066
6	2.063	0.306	1.843	0.268	1.447	0.186	1.127	0.119	0.904	0.079

$n=30$										
d/σ										
2	0.684	0.092	0.611	0.080	0.481	0.056	0.375	0.036	0.301	0.024
3	1.026	0.138	0.917	0.121	0.721	0.084	0.562	0.054	0.451	0.036
4	1.368	0.184	1.222	0.161	0.961	0.112	0.750	0.072	0.602	0.048
5	1.710	0.229	0.528	0.201	1.201	0.140	0.937	0.090	0.752	0.060
6	2.052	0.275	1.833	0.241	1.442	0.167	1.124	0.108	0.903	0.071

Table 3.1 (*Cont.*)

						$|\xi - T|/\sigma$				
	0		0.5		1		1.5		2	
	E	S.D.	E	S.D.	E	S.D.	E	S.D.	E	S.D.
$n=35$										
d/σ										
2	0.681	0.084	0.609	0.074	0.479	0.051	0.374	0.033	0.300	0.022
3	1.022	0.126	0.913	0.111	0.719	0.077	0.561	0.050	0.451	0.033
4	1.363	0.168	1.218	0.148	0.958	0.102	0.748	0.066	0.601	0.044
5	1.703	0.210	1.522	0.184	1.198	0.128	0.935	0.083	0.751	0.055
6	2.044	0.253	1.827	0.221	1.438	0.154	1.122	0.099	0.901	0.066
$n=40$										
d/σ										
2	0.680	0.078	0.607	0.069	0.478	0.048	0.373	0.031	0.300	0.020
3	1.019	0.117	0.911	0.103	0.717	0.071	0.560	0.046	0.450	0.031
4	1.359	0.156	1.215	0.137	0.956	0.095	0.747	0.062	0.600	0.041
5	1.699	0.195	1.518	0.171	1.196	0.119	0.934	0.077	0.750	0.051
6	2.039	0.235	1.822	0.206	1.435	0.143	1.120	0.092	0.901	0.061

$n = 45$

d/σ										
2	0.678	0.073	0.606	0.064	0.477	0.045	0.373	0.029	0.300	0.019
3	1.017	0.110	0.909	0.096	0.716	0.067	0.560	0.043	0.450	0.029
4	1.356	0.147	1.212	0.129	0.955	0.089	0.746	0.058	0.600	0.038
5	1.695	0.183	1.515	0.161	1.194	0.112	0.933	0.072	0.750	0.048
6	2.034	0.220	1.818	0.193	1.432	0.134	1.119	0.087	0.900	0.058

$n = 50$

d/σ										
2	0.677	0.069	0.605	0.061	0.477	0.042	0.373	0.027	0.300	0.018
3	1.015	0.104	0.908	0.091	0.715	0.063	0.559	0.041	0.450	0.027
4	1.354	0.139	1.210	0.121	0.954	0.084	0.745	0.055	0.600	0.036
5	1.692	0.173	1.513	0.152	1.192	0.106	0.932	0.068	0.749	0.046
6	2.031	0.208	1.815	0.182	1.430	0.127	1.118	0.082	0.899	0.055

Table 3.2 Values of C_{pm}

$\dfrac{d}{\sigma}$	\multicolumn{5}{c}{$\dfrac{\|\xi - T\|}{\sigma}$}				
	0.0	0.5	1.0	1.5	2.0
2	0.667	0.596	0.471	0.370	0.298
3	1.000	0.894	0.707	0.555	0.447
4	1.333	1.193	0.943	0.740	0.596
5	1.667	1.491	1.179	0.925	0.745
6	2.000	1.789	1.414	1.109	0.894

is established only under these rather restrictive conditions. If minimization of mean square error were a primary objective, neither n nor $(n-1)$ would be the appropriate divisor in the denominator.

The effect of correlation among the Xs may be less marked for $\hat{C}_{pm}$ than for $\hat{C}_p$ (see section 2.6), because

$$n^{-1}E\left[\sum_{j=1}^{n}(X_j - T)^2\right] = \sigma^2 + (\xi - T)^2$$

whatever the values of $\rho_{ij} = \mathrm{corr}(X_i, X_j)$.

Note on calculation of $E[\tilde{C}_{pm}^2]$

The series for $E[\tilde{C}_{pm}^2]$, obtained by putting $r=2$ in (3.19), is often quite slowly convergent. The following alternative approach was found to give good results.

Since the expected value of $(\chi_{n+2j}^2)^{-1}$ is $(n-2+2j)^{-1}$, the expected value of $\tilde{C}_{pm}^2$ is $(d\sqrt{n}/3\sigma)^2$ times the weighted mean of these quantities with Poisson weights $P_j(\tfrac{1}{2}\lambda)$. We can write

$$E[\tilde{C}_{pm}^2] = \left(\frac{d\sqrt{n}}{3\sigma}\right)^2 E_J\left[\frac{1}{n-2+2J}\right] \qquad (3.24)$$

where J has a Poisson distribution with expected value $\frac{1}{2}\lambda$.

To evaluate $E_J[(n-2+2J)^{-1}]$, we resort to the method of statistical differentials (or 'delta method'), obtaining

$$E\left[\frac{1}{n-2+2J}\right] \cong \frac{1}{n-2+2E[J]}$$

$$+ \sum_{r=1} \frac{d^r(n-2+2J)^{-1}}{dJ^r}\bigg|_{J=E[J]=\lambda/2} \cdot \frac{\mu_r(J)}{r!}$$

(3.25)

The quantities $\mu_r(J)$ are the central moments of the Poisson distribution with expected value $\frac{1}{2}\lambda$, namely

$$E[J]=\tfrac{1}{2}\lambda;\ \mu_2(J)=\tfrac{1}{2}\lambda;\ \mu_3(J)=\tfrac{1}{2}\lambda;\ \mu_4(J)=\tfrac{1}{2}\lambda+\tfrac{3}{4}\lambda^2;\ \cdots$$

(Higher moments are given by Haight (1967, p. 7) or can be evaluated using the recurrence formula – for moments $\mu_r'(J)$ about zero –

$$\mu_{r+1}'(J)=\frac{\lambda}{2}\left[r\mu_{r-1}'(J)+\frac{\partial\mu_r'(J)}{\partial(\lambda/2)}\right]$$

– see e.g. Johnson *et al.* (1993, Ch. 3).)

3.4.2 Confidence interval estimation

Since

$$C_{pm}=\tfrac{1}{3}d\{\sigma^2+(\xi-T)^2\}^{-\frac{1}{2}}$$

construction of a confidence interval for C_{pm} is equivalent to construction of a confidence interval for $\sigma^2+(\xi-T)^2$. If statistics $A(\mathbf{X})$, $B(\mathbf{X})$, based on $X_1, X_2, \ldots, X_n$, could be found, such that

$$\Pr[A(\mathbf{X}) \leqslant \sigma^2 + (\xi - T)^2 \leqslant B(\mathbf{X})] = 1 - \alpha \qquad (3.26)$$

(i.e. $(A(\mathbf{X})$, $B(\mathbf{X}))$ is a $100(1-\alpha)\%$ confidence interval for $\sigma^2 + (\xi - T)^2$) then

$$\left(\frac{d}{3(B(\mathbf{X}))^{\frac{1}{2}}}, \frac{d}{3(A(\mathbf{X}))^{\frac{1}{2}}} \right)$$

would be a $100(1-\alpha)\%$ interval for C_{pm}.

However, it is not possible to find statistics satisfying (3.26) exactly, on the usual assumption of normality of distribution of X. We therefore try to construct approximate $100(1-\alpha)\%$ confidence intervals.

Recalling that $\Sigma_{i=1}^{n}(X_i - T)^2$ is distributed as $\sigma^2 \chi_n'^2(\lambda)$ with $\lambda = n\{(\xi - T)/\sigma\}^2$, we have

$$\text{var}\left(\sum_{i=1}^{n} (X_i - T)^2 \right) = 2n\sigma^2\{\sigma^2 + 2(\xi - T)^2\} \qquad (3.27)$$

(as well as

$$\mathrm{E}\left[\sum_{i=1}^{n} (X_i - T)^2 \right] = \sigma^2 + (\xi - T)^2$$

which is valid if $\mathrm{E}[X] = \xi$ and $\text{var}(X) = \sigma^2$, whatever the distribution of X).

If we use a normal approximation to the distribution of

$$\sum_{i=1}^{n} (X_i - T)^2$$

for large n we obtain the approximate $100(1-\alpha)\%$ interval

$$\frac{1}{n}\sum_{i=1}^{n}(X_i-T)^2\pm z_{1-\alpha/2}\,\sigma[2n^{-1}\{\sigma^2+2(\xi-T)^2\}]^{\frac{1}{2}} \quad (3.28)$$

where $\Phi(z_{1-\alpha/2})=1-\alpha/2$, for $\sigma^2+(\xi-T)^2$. However since ξ and σ are not known (otherwise we would know the value of C_{pm} and not need to estimate it), it is necessary to replace them by estimates. The resulting interval for C_{pm} is of quite complicated form.

Boyles (1991) and Subbaiah and Taam (1991) use a better approximation to the distribution of $\chi_n'^{2}(\lambda)$, namely that introduced by Patnaik (1949). This is the distribution of $c\chi_f^2$ where

$$c=\frac{\sigma^2+2(\xi-T)^2}{\sigma^2+(\xi-T)^2}\quad\text{and}\quad f=\frac{\{\sigma^2+(\xi-T)^2\}^2}{\sigma^2+2(\xi-T)^2}\frac{n}{\sigma^2}$$

(The values of c and f give the correct expected value and variance.)

It is still necessary to use estimates $\hat{c}$ and $\hat{f}$, of c and f, respectively.

Now

$$\left(\frac{C_{pm}}{\tilde{C}_{pm}}\right)^2=\frac{\dfrac{1}{n}\sum_{i=1}^{n}(X_i-T)^2}{\sigma^2+(\xi-T)^2}=\frac{\dfrac{1}{n}\sigma^{-2}\sum_{i=1}^{n}(X_i-T)^2}{1+\left\{\dfrac{\xi-T}{\sigma}\right\}^2}$$

$$=\frac{\sigma^{-2}\sum_{i=1}^{n}(X_i-T)^2}{cf} \quad (3.29)$$

and hence is approximately distributed as

$$\frac{\sigma^2 c \chi_f^2}{\sigma^2 c f} = \frac{\chi_f^2}{f}$$ (3.30)

and

$$\Pr[(f^{-1}\chi_{f,\alpha/2}^2)^{\frac{1}{2}}\tilde{C}_{pm} < C_{pm} < (f^{-1}\chi_{f,1-\alpha/2}^2)^{\frac{1}{2}}\tilde{C}_{pm}] \cong 1 - \alpha$$
(3.31)

where $\chi_{f,\varepsilon}^2$ is defined by

$$\Pr[\chi_f^2 < \chi_{f,\varepsilon}^2] = \varepsilon$$

We construct an approximate $100(1-\alpha)\%$ confidence interval

$$\left(\left(\frac{\chi_{f,\alpha/2}^2}{\hat{f}}\right)^{\frac{1}{2}}\tilde{C}_{pm}; \left(\frac{\chi_{f,1-\alpha/2}^2}{f}\right)^{\frac{1}{2}}\tilde{C}_{pm}\right)$$ (3.32)

for C_{pm} by using the estimate

$$\hat{f} = \frac{n\{\hat{\sigma}^2 + (\hat{\xi} - T)^2\}^2}{\hat{\sigma}^2\{\hat{\sigma}^2 + 2(\hat{\xi} - T)^2\}}$$

where

$$\hat{\sigma}^2 = \frac{1}{n-1}\sum_{i=1}^{n}(X_i - \bar{X})^2 \quad \text{and} \quad \hat{\xi} = \bar{X}$$

Further approximating the distribution of χ_f^2 leads to the approximate $100(1-\alpha)\%$ confidence intervals

$$\tilde{C}_{pm}\left(1 \pm \frac{z_{1-\alpha/2}}{\sqrt{2\hat{f}}}\right)$$ (3.33)

and

$$\tilde{C}_{pm}\left(1-\frac{2}{9\hat{f}}\pm z_{1-\alpha/2}\left(\frac{2}{9\hat{f}}\right)^{\frac{1}{2}}\right)^{\frac{3}{2}} \qquad (3.34)$$

For practical use, we recommend (3.33).

Several other approximate formulas can be obtained for $100(1-\alpha)\%$ confidence intervals for C_{pm}, using approximate distributions of various functions of $\tilde{C}_{pm}$ (e.g. $\log \tilde{C}_{pm}$, $\sqrt{\tilde{C}_{pm}}$, etc.). Subbaiah and Taam (1991) present results of a comparative Monte Carlo study of sixteen such formulas.

From (3.17) it follows that $(C_{pm}/\tilde{C}_{pm})^2$ is distributed as

$$\frac{1}{n}\left\{1+\left(\frac{\xi-T}{\sigma}\right)^2\right\}^{-1} \times \begin{array}{l} \text{[noncentral } \chi^2 \text{ with } n \text{ degrees of} \\ \text{freedom and noncentrality} \\ \text{parameter } n\{(\xi-T)/\sigma\}^2] \end{array}$$

Writing $(\xi-T)/\sigma=\delta$, we have

$$\Pr\left[\chi'^2_{n,\,\alpha/2}(n\delta^2)<n(1+\delta^2)\left(\frac{C_{pm}}{\tilde{C}_{pm}}\right)^2<\chi'^2_{n,\,1-\alpha/2}(n\delta^2)\right]=1-\alpha$$

whence

$$\Pr\left[\left(\frac{1}{n(1+\delta^2)}\right)^{\frac{1}{2}}\chi'_{n,\,\alpha/2}(n\delta^2)\tilde{C}_{pm}<C_{pm}<\right.$$

$$\left.<\left(\frac{1}{n(1+\delta^2)}\right)^{\frac{1}{2}}\chi'_{n,\,1-\alpha/2}(n\delta^2)\tilde{C}_{pm}\right]=1-\alpha$$

It is tempting to say that

$$\left(\left(\frac{1}{n(1+\delta^2)}\right)^{\frac{1}{2}}\chi'_{n,\,\alpha/2}(n\delta^2)\hat{C}_{pm},\left(\frac{1}{n(1+\delta^2)}\right)^{\frac{1}{2}}\chi'_{n,\,1-\alpha/2}(n\delta^2)\hat{C}_{pm}\right)$$

provides a $100(1-\alpha)\%$ confidence interval for C_{pm}. Unfortunately, calculation of these confidence limits involves knowledge of the value of δ. If the natural estimator

$$\hat{\delta} = \frac{\bar{X} - T}{S}$$

is used, the resulting confidence interval will not have $100(1-\alpha)\%$ coverage, but something rather less.

If $\xi = T$, then $C_{\text{pm}} = C_{\text{p}}$ (see 3.21 a), and $\tilde{C}_{\text{pm}}$ is distributed as

$$\left(\frac{d}{3\sigma}\right)\frac{\sqrt{n}}{\chi_n} = \frac{\sqrt{n}}{\chi_n}C_{\text{pm}} = \frac{\sqrt{n}}{\chi_n}C_{\text{p}}$$

The distribution of $\tilde{C}_{\text{pm}}$ for sample size n is thus the same as that of $\hat{C}_{\text{p}}$ for sample size $(n+1)$. The unit increase in effective size of sample comes from the additional knowledge that $\xi = T$. In particular, we obtain $100(1-\alpha)\%$ confidence limits

$$\left(\frac{\chi_{n,\,\alpha/2}}{\sqrt{n}}\tilde{C}_{\text{pm}} \cdot \frac{\chi_{n,\,1-\alpha/2}}{\sqrt{n}}\tilde{C}_{\text{pm}}\right)$$

for $C_{\text{pm}}(=C_{\text{p}})$ in this case.

Chan *et al.* (1988 a) discuss the use of Bayesian methods (with a 'noninformative' prior distribution for σ) in this situation, and provide some tables. See also Cheng (1992) for a nontechnical description (with further tables) of this approach.

3.5 BOYLES' MODIFIED C_{pm} INDEX

Boyles (1992) has proposed the PCI

$$C_{\text{pm}}^{+} = \tfrac{1}{3}[(T-\text{LSL})^{-2}\text{E}_{X<T}[(X-T)^2] \\ + (\text{USL}-T)^{-2}\text{E}_{X>T}[(X-T)^2]]^{-\frac{1}{2}} \quad (3.35)$$

as a modification of C_{pm} for use when $\xi \neq T$. Here

$$E_{X<T}[(X-T)^2] = E[(X-T)^2 | X<T]\Pr[X<T] \tag{3.36 a}$$

and

$$E_{X>T}[(X-T)^2] = E[(X-T)^2 | X>T]\Pr[X>T] \tag{3.36 b}$$

These are sometimes called *semivariances*. (The possibility that X might equal T is ignored, since we suppose that we are dealing with a continuous variable, and so $\Pr[X=T]=0$.)

Boyles introduced this index from a loss function point of view, to allow for an asymmetric loss function. However, it has some features in common with the index C_{jkp}, to be introduced in Section 4.4 which is intended to allow for some asymmetry in the distribution of X (the process distribution).

Note that, if $T=\frac{1}{2}(\text{LSL}+\text{USL})$, then $C_{pm}^+ = C_{pm}$, so C_{pm}^+ is not affected by asymmetry of the process distribution in that case.

A natural estimator of C_{pm}^+ is

$$\hat{C}_{pm}^+ = \frac{1}{3}\left[\frac{1}{n}\left\{(T-\text{LSL})^{-2}\sum_{X_i<T}(X_i-T)^2\right.\right.$$
$$\left.\left. + (\text{USL}-T)^{-2}\sum_{X_i>T}(X_i-T)^2\right\}\right]^{-\frac{1}{2}} \tag{3.37}$$

In general, the distribution of $\hat{C}_{pm}^+$ is complicated, even when the process distribution is normal.

We note that

$$W = \frac{n}{9\hat{C}_{pm}^{+2}} = (T-\text{LSL})^{-2}\sum_{X_i<T}(X_i-T)^2$$
$$+ (\text{USL}-T)^{-2}\sum_{X_i>T}(X_i-T)^2 \tag{3.38}$$

We now derive expressions for the expected value and variance of W when X_i is distributed normally with expected value ξ equal to T, and variance σ^2. (Of course, ξ and σ are not in fact known; otherwise we woud not need to estimate them. The analysis studies what would happen if $\xi = T = m$ and the formula for $\hat{C}_{pm}^{+}$ is used.) Conditionally on there being K Xs less than T and $(n-K)$ greater than T, W is distributed as

$$\{(T-\mathrm{LSL})^{-2}\chi_K^2 + (\mathrm{USL} - T)^{-2}\chi_{n-K}^2\}\sigma^2 \qquad (3.39)$$

the two χ^2s being mutually independent. The conditional expected value and variance of W, given K are

$$\mathrm{E}[W|K] = \{(T-\mathrm{LSL})^{-2}K + (\mathrm{USL} - T)^{-2}(n-K)\}\sigma^2$$
$$(3.40\,a)$$

and

$$\mathrm{var}[W|K] = 2\{(T-\mathrm{LSL})^{-4}K + (\mathrm{USL} - T)^{-4}(n-K)\}\sigma^4$$
$$(3.40\,b)$$

Since $\Pr[X = T] = 0$, the distribution of K is binomial with parameters $(n, \frac{1}{2})$, hence

$$\mathrm{E}[K] = \mathrm{E}[n-K] = \tfrac{1}{2}n \qquad (3.41\,a)$$

$$\mathrm{var}(K) = \mathrm{var}(n-K) = \tfrac{1}{4}n \qquad (3.41\,b)$$

$$\mathrm{E}[K^2] = \mathrm{var}(K) + \{\mathrm{E}[K]\}^2 = \tfrac{1}{4}n(n+1) = \mathrm{E}[(n-K)^2]$$
$$(3.14\,c)$$

and

$$\mathrm{E}[K(n-K)] = n\mathrm{E}[K] - \mathrm{E}[K^2] = \tfrac{1}{4}n(n-1) \qquad (3.41\,d)$$

Hence the overall expected value of W is

$$\mathrm{E}[W] = \mathrm{E}[\mathrm{E}[W\,|\,K]] = \tfrac{1}{2}n\{(T-\mathrm{LSL})^{-2}+(\mathrm{USL}-T)^{-2}\}\sigma^2$$
$$(3.42\,a)$$

and the overall variance is

$$
\begin{aligned}
\mathrm{var}(W) &= \mathrm{E}[\mathrm{E}[W^2|K]] - \{\mathrm{E}[W]\}^2 \\
&= \mathrm{E}[2\{(T-\mathrm{LSL})^{-4}K \\
&\quad + (\mathrm{USL}-T)^{-4}(n-K)\} + \{(T-\mathrm{LSL})^{-2}K \\
&\quad + (\mathrm{USL}-T)^{-2}(n-K)\}^2] - \tfrac{1}{4}n^2\{(T-\mathrm{LSL})^{-2} \\
&\quad + (\mathrm{USL}-T)^{-2}\}^2\sigma^4 \\
&= n[(T-\mathrm{LSL})^{-4}+(\mathrm{USL}-T)^{-4}+\tfrac{1}{4}\{(T-\mathrm{LSL})^{-2} \\
&\quad - (\mathrm{USL}-T)^{-2}\}^2]\sigma^4 \qquad\qquad (3.42\,b)
\end{aligned}
$$

Since X is distributed as $\mathrm{N}(T,\sigma^2)$ we have $\xi = T$. Hence

$$C_{\text{pm}}^{+2} = \frac{1}{9 \times \tfrac{1}{2}\{(T-\mathrm{LSL})^{-2}+(\mathrm{USL}-T)^{-2}\}\sigma^2}$$

and so

$$\frac{C_{\text{pm}}^{+2}}{\hat{C}_{\text{pm}}^{+2}} = \frac{W}{\tfrac{1}{2}n\{(T-\mathrm{LSL})^{-2}+(\mathrm{USL}-T)^{-2}\}\sigma^2} \qquad (3.43)$$

whence

$$\mathrm{E}\!\left[\left(\frac{C_{\text{pm}}^{+}}{\hat{C}_{\text{pm}}^{+}}\right)^2\right] = 1 \qquad (3.44\,a)$$

and

$$\text{var}\left(\left(\frac{C^+_{pm}}{\hat{C}^+_{pm}}\right)^2\right) = \frac{\text{var}(W)}{[\frac{1}{2}n\{(T-LSL)^{-2}+(USL-T)^{-2}\}\sigma^2]^2}$$

$$= \frac{4\{(T-LSL)^{-4}+(USL-T)^{-4}\}+\{(T-LSL)^{-2}-(USL-T)^{-2}\}^2}{\{(T-LSL)^{-2}+(USL-T)^{-2}\}^2}$$

$$(3.44\,b)$$

Boyles (1992) suggests that the distribution of $(C^+_{pm}/\hat{C}^+_{pm})^2$ might be well approximated by that of a 'mean chi-squared' – χ^2_v/v. This would give the correct first two moments, if v has the value

$$\frac{n\{(T-LSL)^{-2}+(USL-T)^{-2}\}^2}{2\{(T-LSL)^{-4}+(USL-T)^{-4}\}+\frac{1}{2}\{(T-LSL)^{-2}-(USL-T)^{-2}\}^2}$$

$$= \frac{n(1+R^2)^2}{2(1+R^4)+\frac{1}{2}(1-R^2)^2}$$

with

$$R = \frac{USL-T}{T-LSL} \tag{3.45}$$

[If $T=\frac{1}{2}(LSL+USL)$, then $R=1$.]

Although $\hat{C}^{+2}_{pm}$ is not an unbiased estimator of C^{+2}_{pm} (nor is $\hat{C}^+_{pm}$, of C^+_{pm}), the statistic $W=n/(9\hat{C}^{+2}_{pm})$ is an unbiased estimator of $(9C^{+2}_{pm})^{-1}$. It is not, however, a minimum variance unbiased estimator.

Details of calculation of such an estimator, based on the method outlined in section 1.12.4, are shown in the Appendix to the present chapter.

The index C_{jkp}, to be described in section 4.4 also uses the semivariances, and, in its estimator, the partial sums of squares

$$\sum_{X_i<T}(X_i-T)^2 \quad \text{and} \quad \sum_{X_i>T}(X_i-T)^2.$$

3.6 THE C_{pmk} INDEX

C_{pk} is obtained from C_p by modifying the numerator; C_{pm} is obtained from C_p by modifying the denominator. If the two modifications are combined, we obtain the index

$$C_{pmk} = \frac{d - |\xi - m|}{3\{E[(X-T)^2]\}^{\frac{1}{2}}} = \frac{d - |\xi - m|}{3\{\sigma^2 + (\xi - T)^2\}^{\frac{1}{2}}} \quad (3.46)$$

introduced by Pearn *et al.* (1992). We note that

$$C_{pmk} = \left(1 - \frac{|\xi - m|}{d}\right) C_{pm}$$

$$= \left\{1 + \left(\frac{\xi - T}{\sigma}\right)^2\right\}^{-\frac{1}{2}} C_{pk}$$

$$= \left(1 - \frac{|\xi - m|}{d}\right)\left\{1 + \left(\frac{\xi - T}{\sigma}\right)^2\right\}^{\frac{1}{2}} C_p \quad (3.47)$$

Hence $C_{pmk} \leqslant C_{pm}$, and also $C_{pmk} \leqslant C_{pk}$. In fact $C_{pmk} = C_{pk}C_{pm}/C_p$. If $\xi = m = T$ then $C_p = C_{pk} = C_{pm} = C_{pmk}$. However, if we were estimating C_{pmk} (or the other PCIs) we would not know that $\xi = T = m$.

A natural estimator for C_{pmk} is

$$\tilde{C}_{pmk} = \frac{d - |\bar{X} - m|}{3\left\{\dfrac{1}{n}\displaystyle\sum_{i=1}^{n}(X_i - T)^2\right\}^{\frac{1}{2}}} \quad (3.48)$$

The distribution of $\tilde{C}_{pmk}$ is quite complicated, even if $m = T$. We first suppose, in addition to normality, that $\xi = m = T$ and obtain an expression for the rth moment about zero of $\tilde{C}_{pmk}$ in that case.

If $\xi = T = m$, we have

$$\tilde{C}_{pmk}^r = \frac{n^{\frac{1}{2}r}(d - |\bar{X} - m|)^r}{3^r \left\{ \sum_{i=1}^{n} (X_i - \bar{X})^2 + n(\bar{X} - m)^2 \right\}^{\frac{1}{2}r}}$$

$$= \frac{1}{3^r} \sum_{j=0}^{r} (-1)^j \binom{r}{j} d^{r-j} n^{\frac{1}{2}(r-j)}$$

$$\left\{ \frac{n(\bar{X} - m)^2}{\sum_{i=1}^{n} (X_i - \bar{X})^2 + n(\bar{X} - m)^2} \right\}^{\frac{1}{2}j}$$

$$\times \left\{ \frac{1}{\sum_{i=1}^{n} (X_i - \bar{X})^2 + n(\bar{X} - m)^2} \right\}^{\frac{1}{2}(r-j)} \tag{3.49}$$

We now use the following well known results:

(a) $Y_1 = \sum_{i=1}^{n} (X_i - \bar{X})^2$ is distributed as $\sigma^2 \chi_{n-1}^2$;

(b) $Y_2 = n(\bar{X} - m)^2$ is distributed as $\sigma^2 \chi_1^2$;

(c) $Y_2 / (Y_1 + Y_2)$ has a Beta$(\frac{1}{2}, \frac{1}{2}(n-1))$ distribution (see section 1.6);

(d) $Y_2 / (Y_1 + Y_2)$ and $(Y_1 + Y_2)$ are mutually independent (see section 1.6.)

Then, from (3.49) and (a)–(d)

$$E[\tilde{C}_{pmk}^r] = \frac{1}{3^r} \sum_{j=0}^{r} (-1)^j \binom{r}{j} \left(\frac{d\sqrt{n}}{\sigma} \right)^{r-j}$$

$$\times E[\{Beta(\tfrac{1}{2}, \tfrac{1}{2}(n-1))\}^{\frac{1}{2}j}] \ E[(\chi_n^2)^{-\frac{1}{2}(r-j)}]$$

$$= \frac{1}{3^r \sqrt{\pi}} \sum_{j=0}^{r} (-1)^j \binom{r}{j} \left(\frac{d}{\sigma}\sqrt{\frac{n}{2}}\right)^{r-j}$$

$$\times \frac{\Gamma(\frac{1}{2}(j+1))\Gamma(\frac{1}{2}(n-r+j))}{\Gamma(\frac{1}{2}(n+j))} \qquad (3.50\,a)$$

If we allow for the possibility that ξ is not equal to m (though still assuming $T=m$), then (a) remains unchanged, but Y_2 is distributed as

$\sigma^2 \times$ (noncentral χ^2 with one degree of freedom and
noncentrality parameter $\lambda = \sigma^{-2}(\xi-m)^2) \equiv \sigma^2 \chi_1'^2(\lambda)$

Recalling from (1.33) that $\chi_1'^2(\lambda)$ is distributed as a mixture of central χ_{1+2i}^2 distributions with Poisson weights $\exp(-\frac{1}{2}\lambda)(\frac{1}{2}\lambda)^i/i!$ $(i=0,1,\dots)$, (c) and (d) must be replaced by

(c, d)′ the joint distribution of $Y_2/(Y_1 + Y_2)$ and $(Y_1 + Y_2)$ is
that of a mixture of pairs of independent variables
with distributions Beta$(\frac{1}{2}+i, \frac{1}{2}(n-1))$, χ_{n+2i}^2 respectively and the Poisson weights given above. (Remember
that $\chi_{n+2i}'^2(\lambda)+\chi_{n-1}^2$ is distributed as $\chi_{n+2i}'^2(\lambda)$ if the
two variates are independent.)

For this more general case we find

$$E[\tilde{C}_{\text{pmk}}^r] = 3^{-r} \exp(-\lambda/2) \sum_{i=0}^{\infty} \frac{(\lambda/2)^i}{i!} \sum_{j=0}^{r} (-1)^j \binom{r}{j} \left(\frac{d\sqrt{n}}{\sigma}\right)^{r-j}$$

$$\times E[\{\text{Beta}(\tfrac{1}{2}+i, \tfrac{1}{2}(n-1))\}^{j/2}] E[(\chi_{1+2i}^2)^{-(r-j)/2}]$$

$$= 3^{-r} \exp(-\lambda/2) \sum_{i=0}^{\infty} \frac{(\lambda/2)^i}{i!}$$

$$\times \sum_{j=0}^{r} (-1)^j \binom{r}{j} \left(\frac{d}{\sigma}\sqrt{\frac{n}{2}}\right)^{r-j}$$

$$\times \frac{\Gamma(\frac{1}{2}(1+j)+i)\Gamma(\frac{1}{2}(n-r+j)+i)}{\Gamma(\frac{1}{2}+i)\Gamma(\frac{1}{2}(n+j)+i)} \tag{3.50 b}$$

Taking $r=1$ we obtain

$$E[\tilde{C}_{pmk}] = \frac{1}{3}\exp(-\lambda/2) \sum_{i=0}^{\infty} \frac{(\lambda/2)^i}{i!} \left[\frac{d}{\sigma}\sqrt{\frac{n}{2}}\frac{\Gamma(\frac{1}{2}(n-1)+i)}{\Gamma(\frac{1}{2}n+i)}\right.$$

$$\left. - \frac{\Gamma(1+i)\Gamma(\frac{1}{2}n+i)}{\Gamma(\frac{1}{2}+i)\Gamma(\frac{1}{2}(n+1)+i)}\right]$$

$$= \frac{1}{3}\exp(-\lambda/2) \sum_{i=0}^{\infty} \frac{(\lambda/2)^i}{i!} \left[\frac{d}{\sigma}\sqrt{\frac{n}{2}}R_i\right.$$

$$\left. - \frac{2\Gamma(1+i)}{(n-1+2i)\Gamma(\frac{1}{2}+i)}\frac{1}{R_i}\right] \tag{3.50 c}$$

with $R_i = \Gamma(\frac{1}{2}(n-1)+i)/\Gamma(\frac{1}{2}n+i)$. Taking $\lambda=0$ we find

$$E[\tilde{C}_{pmk}] = \frac{1}{3}\left[\frac{d}{\sigma}\sqrt{\frac{n}{2}}R - \frac{2}{(n-1)\sqrt{\pi}}\frac{1}{R}\right] \tag{3.50 d}$$

with $R = \Gamma(\frac{1}{2}(n-1))/\Gamma(\frac{1}{2}n)$.

Now taking $r=2$, we obtain from (3.50 b) after some simplification

$$E[\tilde{C}_{pmk}^2] = \frac{1}{9}\exp(-\frac{1}{2}\lambda) \sum_{i=0}^{\infty} \frac{(\frac{1}{2}\lambda)^i}{i!} \left[\left(\frac{d\sqrt{n}}{\sigma}\right)^2 \frac{1}{n-2+2i}\right.$$

$$\left. -2\sqrt{2}\left(\frac{d\sqrt{n}}{\sigma}\right)\frac{\Gamma(1+i)}{\Gamma(\frac{1}{2}+i)}\left(\frac{1}{n-1+2i}\right) + \frac{1+2i}{n+2i}\right] \tag{3.50 e}$$

From $(3.50\,c)$ and $(3.50\,d)$, the variance of $\tilde{C}_{pmk}$ can be derived. If $\lambda = 0$ we have

$$E[\tilde{C}_{pmk}^2] = \frac{1}{9}\left[\frac{1}{n-2}\left(\frac{d\sqrt{n}}{\sigma}\right)^2 - \frac{2\sqrt{2}}{(n-1)\sqrt{\pi}}\left(\frac{d\sqrt{n}}{\sigma}\right) + \frac{1}{n}\right]$$

$$(3.50f)$$

Table 3.3 gives values of expected values and standard deviations of $\tilde{C}_{pmk}$, from (3.50) for a few values of n and d/σ. Remember that these values apply for $\xi = T = m$.

Just as $C_{pmk} \leqslant C_{pm}$, with equality if $\xi = m$, a similar argument shows that $\tilde{C}_{pmk} \leqslant \tilde{C}_{pm}$. However $\mathrm{var}(\tilde{C}_{pmk}) \geqslant \mathrm{var}(\tilde{C}_{pm})$. If $T = m$ we have from (3.46)

$$C_{pmk} = \frac{d}{3\sigma}\frac{1 - \left|\dfrac{\xi - m}{\sigma}\right|\dfrac{\sigma}{d}}{\left\{1 + \left(\dfrac{\xi - m}{\sigma}\right)^2\right\}^{\frac{1}{2}}} \qquad (3.51)$$

Table 3.4 gives values of C_{pmk} for representative values of d/σ and $|\xi - m|/\sigma$.

3.7 $C_{pm}(a)$ INDICES

The index C_{pm} attempts to represent a combination of effects of greater (or less) variability and of greater (or less) relative deviation of ξ from the target value. The class of indices now to be described attempts to provide a flexible choice of the relative importance to be ascribed to these two effects.

From (3.8), we note that if $|\zeta| = |\xi - T|/\sigma$ is small, then

$$C_{pm} \cong (1 - \tfrac{1}{2}\zeta^2)C_p \qquad (3.52)$$

Table 3.3 Moments of $\hat{C}_{pmk}$: $E = E[\tilde{C}_{pmk}]$; S.D. $=$ S.D. $(\tilde{C}_{pmk})$

| | $|\xi - m|/\sigma$ | | | | | | | | | |
| | 0.0 | | 0.5 | | 1.0 | | 1.5 | | 2.0 | |
	E	S.D.	E	S.D.	E	S.D.	E	S.D.	E	S.D.
$n=10$										
d/σ										
2	0.636	0.193	0.491	0.192	0.264	0.135	0.106	0.081	0.006	0.051
3	0.997	0.281	0.813	0.268	0.515	0.187	0.299	0.114	0.160	0.072
4	1.359	0.371	1.126	0.347	0.765	0.241	0.492	0.148	0.313	0.094
5	1.720	0.462	1.458	0.426	1.016	0.295	0.684	0.182	0.467	0.116
6	2.081	0.553	1.780	0.505	1.266	0.350	0.877	0.216	0.620	0.138
$n=20$										
d/σ										
2	0.633	0.124	0.470	0.129	0.249	0.087	0.099	0.054	0.003	0.035
3	0.979	0.180	0.779	0.176	0.492	0.121	0.288	0.076	0.154	0.049
4	1.326	0.237	1.089	0.225	0.734	0.155	0.476	0.098	0.305	0.063
5	1.672	0.294	1.398	0.275	0.977	0.190	0.665	0.120	0.457	0.078
6	2.019	0.352	1.708	0.325	1.220	0.225	0.854	0.142	0.608	0.093

n = 30

d/σ										
2	0.635	0.099	0.462	0.103	0.244	0.070	0.097	0.043	0.002	0.028
3	0.977	0.142	0.768	0.141	0.485	0.096	0.284	0.060	0.152	0.039
4	1.319	0.187	1.074	0.179	0.725	0.123	0.471	0.078	0.303	0.051
5	1.661	0.232	1.379	0.218	0.965	0.151	0.659	0.096	0.453	0.063
6	2.003	0.278	1.685	0.258	1.206	0.178	0.846	0.114	0.604	0.075

n = 40

d/σ										
2	0.637	0.084	0.459	0.088	0.242	0.060	0.096	0.037	0.002	0.024
3	0.977	0.121	0.762	0.120	0.481	0.082	0.282	0.052	0.152	0.034
4	1.317	0.160	1.066	0.153	0.720	0.105	0.469	0.067	0.302	0.044
5	1.656	0.198	1.370	0.186	0.959	0.129	0.656	0.082	0.452	0.054
6	1.996	0.237	1.673	0.220	1.199	0.152	0.843	0.097	0.602	0.064

n = 50

d/σ										
2	0.639	0.075	0.456	0.079	0.241	0.053	0.095	0.033	0.001	0.021
3	0.978	0.108	0.759	0.107	0.479	0.073	0.281	0.046	0.151	0.030
4	1.316	0.141	1.061	0.136	0.718	0.093	0.468	0.059	0.301	0.039
5	1.654	0.175	1.364	0.165	0.956	0.114	0.654	0.073	0.451	0.048
6	1.993	0.210	1.666	0.195	1.194	0.135	0.840	0.087	0.601	0.057

Table 3.3 (*Cont.*)

					$|\xi - m|/\sigma$					
	0.0		0.5		1.0		1.5		2.0	
	E	S.D.	E	S.D.	E	S.D.	E	S.D.	E	S.D.
$n=60$										
d/σ										
2	0.641	0.068	0.455	0.071	0.240	0.048	0.095	0.030	0.001	0.019
3	0.978	0.098	0.756	0.097	0.478	0.066	0.281	0.042	0.151	0.027
4	1.316	0.128	1.058	0.123	0.716	0.085	0.467	0.054	0.301	0.036
5	1.653	0.159	1.360	0.150	0.954	0.104	0.653	0.066	0.450	0.044
6	1.991	0.190	1.662	0.177	1.192	0.123	0.830	0.079	0.600	0.052
$n=70$										
d/σ										
2	0.642	0.063	0.454	0.066	0.239	0.044	0.094	0.028	0.001	0.018
3	0.979	0.090	0.755	0.089	0.477	0.061	0.280	0.039	0.151	0.025
4	1.316	0.118	1.056	0.114	0.715	0.078	0.466	0.050	0.300	0.033
5	1.653	0.147	1.357	0.138	0.952	0.096	0.652	0.061	0.450	0.040
6	1.990	0.175	1.659	0.163	1.190	0.113	0.838	0.073	0.599	0.048

n = 80

d/σ										
2	0.643	0.058	0.453	0.061	0.239	0.041	0.094	0.026	0.001	0.017
3	0.980	0.084	0.754	0.083	0.476	0.057	0.280	0.036	0.150	0.024
4	1.316	0.110	1.055	0.106	0.714	0.073	0.466	0.047	0.300	0.031
5	1.653	0.137	1.355	0.129	0.951	0.089	0.651	0.057	0.449	0.038
6	1.989	0.163	1.656	0.152	1.188	0.105	0.837	0.068	0.599	0.045

n = 90

d/σ										
2	0.644	0.055	0.452	0.058	0.239	0.039	0.094	0.024	0.001	0.016
3	0.980	0.079	0.753	0.078	0.476	0.054	0.280	0.034	0.150	0.022
4	1.316	0.104	1.053	0.100	0.713	0.069	0.465	0.044	0.300	0.029
5	1.653	0.129	1.354	0.121	0.950	0.084	0.651	0.054	0.449	0.036
6	1.989	0.154	1.654	0.143	1.187	0.099	0.837	0.064	0.599	0.042

n = 100

d/σ										
2	0.645	0.052	0.452	0.055	0.238	0.037	0.094	0.023	0.001	0.015
3	0.981	0.075	0.752	0.074	0.475	0.051	0.279	0.032	0.150	0.021
4	1.317	0.098	1.052	0.094	0.712	0.065	0.465	0.042	0.300	0.027
5	1.653	0.122	1.353	0.115	0.949	0.079	0.651	0.051	0.449	0.034
6	1.988	0.146	1.653	0.135	1.186	0.094	0.836	0.060	0.599	0.040

Table 3.4 Values of C_{pmk} when $T=m$

$\dfrac{d}{\sigma}$	$\dfrac{\lvert\xi-m\rvert}{\sigma}$				
	0.0	0.5	1.0	1.5	2.0
2	0.667	0.447	0.236	0.0925	0.000
3	1.000	0.745	0.471	0.277	0.149
4	1.333	1.043	0.707	0.462	0.298
5	1.667	1.342	1.943	0.647	0.447
6	2.000	1.640	1.179	1.832	0.596

It is suggested that we consider the class of PCIs defined by

$$C_{pm}(a)=(1-a\zeta^2)C_p \qquad (3.53)$$

where a is a positive constant, chosen with regard to the desired balance between the effects of variability and relative departure from target value.

A natural estimator of $C_{pm}(a)$ is

$$\hat{C}_{pm}(a)=\left\{1-a\left(\frac{\bar{X}-T}{S}\right)^2\right\}\hat{C}_p=\frac{d}{3S}\left\{1-a\left(\frac{\bar{X}-T}{S}\right)^2\right\} \qquad (3.54)$$

Assuming normal $(N(\xi,\sigma^2))$ variation, $\hat{C}_{pm}(a)$ is distributed as

$$\frac{(n-1)^{\frac{1}{2}}}{\chi_{n-1}}\left\{1-\frac{a(n-1)}{n\chi^2_{n-1}}(U+\zeta\sqrt{n})^2\right\}C_p \qquad (3.55)$$

where U and χ_{n-1} are mutually independent and U is a unit normal variable.

From (3.55), the r-th moment of $\hat{C}_{pm}(a)$ is

$$E[\{\hat{C}_{pm}(a)\}^r]=$$

$$= (n-1)^{r/2} \sum_{j=0}^{r} (-1)^j \binom{r}{j} \left\{ \frac{a(n-1)}{n} \right\}^j - E[(U + \xi\sqrt{n})^{2j}] E[\chi_{n-1}^{-r-2j}]$$

(3.56 a)

In particular using (2.16 a, b)

$$E[\hat{C}_{pm}(a)] = \left\{ 1 - \frac{a(n-1)}{n(n-4)} (1 + n\zeta^2) \right\} E[\hat{C}_p]$$

(3.56 b)

and

$$E[\{\hat{C}_{pm}(a)\}^2] = \left\{ 1 - \frac{2a(n-1)}{n(n-5)}(1 + n\zeta^2) + \right.$$

$$\left. \frac{a^2(n-1)^2}{n^2(n-5)(n-7)} (3 + 6n\zeta^2 + n^2\zeta^4) \right\} E[\hat{C}_p^2]$$

(3.56 c)

It must be remembered that the $C_{pm}(a)$ indices suffer from the same drawback as that noted for C_{pm} in section 3.2, namely, that if T is not equal to $m = \frac{1}{2}(\text{LSL} + \text{USL})$, the same value of $C_{pm}(a)$, obtained when $\xi = T - \delta$ or $\xi = T + \delta$, can correspond to markedly different expected proportions of NC items.

It is possible for a $C_{pm}(a)$ index to be negative (unlike C_p or C_{pm}, but like C_{pk}). Since this occurs only for large values of $|\zeta|$, the relative departure of ξ from the target value T, this would not be a drawback.

Gupta and Kotz (1993) have tabulated values of $E[\hat{C}_{pm}(\frac{1}{2})]$ and $\text{S.D.}(\hat{C}_{pm}(\frac{1}{2}))$.

APPENDIX 3.A: A MINIMUM VARIANCE UNBIASED ESTIMATOR FOR $W = (9C_{pm}^{+2})^{-1}$

This appendix is for more mathematically-inclined readers.

If the independent random variables $X_1, \ldots, X_n$ each have the normal distribution with expected value ξ and variance

σ^2, then

$$\frac{\left\{\left(\frac{n}{n-1}\right)^{\frac{1}{2}}(X_i - \bar{X})\right\}^2}{\sum\limits_{j=1}^{n}(X_j - \bar{X})^2} \quad \text{with } \bar{X} = \frac{1}{n}\sum\limits_{j=1}^{n} X_j$$

has a beta distribution with parameters $\frac{1}{2}, \frac{1}{2}(n-2)$, as defined in section 1.6 (see e.g., David *et al.* (1954).)

The PDF of

$$Y_i = \frac{n}{(n-1)^2}\left(\frac{X_i - \bar{X}}{S}\right)^2 \quad \text{with } (n-1)S^2 = \sum\limits_{j=1}^{n}(X_j - \bar{X})^2$$

is thus

$$f_{Y_i}(y) = \{B(\tfrac{1}{2}, \tfrac{1}{2}n - 1)\}^{-1} y^{-\frac{1}{2}}(1-y)^{\frac{1}{2}n-2} \quad 0 < y < 1 \quad (3.57)$$

and the PDF of

$$V_i = \frac{\sqrt{n}}{n-1}\frac{X_i - \bar{X}}{S} = \sqrt{Y_i}\,\text{sgn}(X_i - \bar{X}))$$

is

$$f_{V_i}(v) = \{B(\tfrac{1}{2}, \tfrac{1}{2}n - 1)\}^{-1}(1-v^2)^{\frac{1}{2}n-2} \quad -1 < v < 1 \quad (3.58)$$

The conditional distribution of X_i, given the sufficient statistics $\bar{X}$ and S, is that of the linear function of V_i:

$$\bar{X} + \frac{n-1}{\sqrt{n}} S V_i$$

Now, the inequality $X_i < T$ is equivalent to

$$V_i < -\frac{\sqrt{n}}{n-1}\frac{\bar{X}-T}{S} \tag{3.59}$$

Hence

$$E_{X_i < T}[(X_i - T)^2 | \bar{X}, S] =$$

$$= \{B(\tfrac{1}{2}, \tfrac{1}{2}n - 1)\}^{-1} \int_{-1}^{-c_n(\bar{X}-T)/S} (\bar{X} - T + c_n^{-1}Sv)^2 (1 - v^2)^{\frac{1}{2}n - 2} \, dv$$

$$= \left(\frac{S}{c_n}\right)^2 g_n\left(\frac{c_n(\bar{X}-T)}{S}\right) \tag{3.60}$$

with $c_n = \sqrt{n}/(n-1)$ and

$$g_n(z) = \{B(\tfrac{1}{2}, \tfrac{1}{2}n - 1)\}^{-1} \int_{-1}^{-z} (z + v)^2 (1 - v^2)^{\frac{1}{2}n - 2} \, dv \tag{3.61 a}$$

Since $\tfrac{1}{2}(V_i + 1)$ has a beta distribution with parameters $(\tfrac{1}{2}n - 1, \tfrac{1}{2}n - 1)$, an alternative form for $g_n(z)$ can be obtained in terms of incomplete beta function ratios (see section 1.6). We find

$$g_n(z) = 4\{h^2 I_h(\tfrac{1}{2}n - 1, \tfrac{1}{2}n - 1) - h I_h(\tfrac{1}{2}n, \tfrac{1}{2}n - 1)$$

$$+ \frac{n}{2n-1} I_h(\tfrac{1}{2}n + 1, \tfrac{1}{2}n - 1)\} \tag{3.61 b}$$

where $h = \tfrac{1}{2}(1 - z)$.

Since $E[V_i] = 0$, $E[V_i^2] = (n-1)^{-1}$ and $X_i - T = (\bar{X} - T) + (n - 1/\sqrt{n})SV_i$

$$E_{X_i < T}[(X_i - T)^2 | \bar{X}, S] + E_{X_i > T}[(X_i - T)^2 | \bar{X}, S]$$

$$= E[(X_i - T)^2 | \bar{X}, S] = (\bar{X} - T)^2 + \frac{n-1}{n} S^2 \tag{3.62}$$

Table 3.5 Function $g_n(x)$

x					n				
	10	20	30	40	50	100	200	600	1000
−1	1.1111	1.0526	1.0345	1.0256	1.0204	1.0101	1.0050	1.0017	1.0010
−0.95	1.0136	0.9551	0.9370	0.9281	0.9229	0.9126	0.9075	0.9042	0.9035
−0.9	0.9211	0.8626	0.8445	0.8356	0.8304	0.8201	0.8150	0.8117	0.8110
−0.85	0.8336	0.7751	0.7570	0.7481	0.7429	0.7326	0.7275	0.7242	0.7235
−0.8	0.7511	0.6926	0.6745	0.6656	0.6604	0.6501	0.6450	0.6417	0.6410
−0.75	0.6736	0.6151	0.5970	0.5881	0.5829	0.5726	0.5675	0.5642	0.5635
−0.7	0.6010	0.5426	0.5245	0.5156	0.5104	0.5001	0.4950	0.4917	0.4910
−0.65	0.5334	0.4751	0.4570	0.4481	0.4429	0.4326	0.4275	0.4242	0.4235
−0.6	0.4707	0.4126	0.3945	0.3856	0.3804	0.3701	0.3650	0.3617	0.3610
−0.55	0.4128	0.3551	0.3370	0.3281	0.3229	0.3126	0.3075	0.3042	0.3035
−0.5	0.3597	0.3025	0.2845	0.2756	0.2704	0.2601	0.2550	0.2517	0.2510
−0.45	0.3112	0.2549	0.2370	0.2281	0.2229	0.2126	0.2075	0.2042	0.2035
−0.4	0.2672	0.2122	0.1944	0.1856	0.1804	0.1701	0.1650	0.1617	0.1610
−0.35	0.2276	0.1743	0.1568	0.1481	0.1429	0.1326	0.1275	0.1242	0.1235
−0.3	0.1922	0.1410	0.1240	0.1155	0.1103	0.1001	0.0950	0.0917	0.0910
−0.25	0.1608	0.1122	0.0960	0.0877	0.0827	0.0726	0.0675	0.0642	0.0635
−0.2	0.1331	0.0877	0.0724	0.0646	0.0599	0.0500	0.0450	0.0417	0.0410
−0.15	0.1090	0.0672	0.0532	0.0460	0.0416	0.0324	0.0275	0.0242	0.0235

−0.1	0.0883	0.0504	0.0379	0.0315	0.0276	0.0193	0.0149	0.0117	0.0110
−0.05	0.0705	0.0369	0.0261	0.0206	0.0173	0.0105	0.0068	0.0041	0.0035
0	0.0556	0.0263	0.0172	0.0128	0.0102	0.0051	0.0025	0.0008	0.0005
0.05	0.0431	0.0182	0.0109	0.0075	0.0056	0.0021	0.0007	0.0001	0.0000
0.1	0.0328	0.0122	0.0066	0.0042	0.0029	0.0008	0.0001	0.0000	0.0000
0.15	0.0246	0.0079	0.0038	0.0021	0.0013	0.0002	0.0000	0.0000	0.0000
0.2	0.0180	0.0049	0.0020	0.0010	0.0006	0.0001	0.0000	0.0000	0.0000
0.25	0.0129	0.0029	0.0010	0.0004	0.0002	0.0000	0.0000	0.0000	0.0000
0.3	0.0089	0.0016	0.0005	0.0002	0.0001	0.0000	0.0000	0.0000	0.0000
0.35	0.0060	0.0009	0.0002	0.0001	0.0000	0.0000	0.0000	0.0000	0.0000
0.4	0.0039	0.0004	0.0001	0.0000	0.0000	0.0000	0.0000	0.0000	0.0000
0.45	0.0024	0.0002	0.0000	0.0000	0.0000	0.0000	0.0000	0.0000	0.0000
0.5	0.0014	0.0001	0.0000	0.0000	0.0000	0.0000	0.0000	0.0000	0.0000
0.55	0.0008	0.0000	0.0000	0.0000	0.0000	0.0000	0.0000	0.0000	0.0000
0.6	0.0004	0.0000	0.0000	0.0000	0.0000	0.0000	0.0000	0.0000	0.0000
0.65	0.0002	0.0000	0.0000	0.0000	0.0000	0.0000	0.0000	0.0000	0.0000
0.7	0.0001	0.0000	0.0000	0.0000	0.0000	0.0000	0.0000	0.0000	0.0000
0.75	0.0000	0.0000	0.0000	0.0000	0.0000	0.0000	0.0000	0.0000	0.0000
0.8	0.0000	0.0000	0.0000	0.0000	0.0000	0.0000	0.0000	0.0000	0.0000
0.85	0.0000	0.0000	0.0000	0.0000	0.0000	0.0000	0.0000	0.0000	0.0000
0.9	0.0000	0.0000	0.0000	0.0000	0.0000	0.0000	0.0000	0.0000	0.0000
0.95	0.0000	0.0000	0.0000	0.0000	0.0000	0.0000	0.0000	0.0000	0.0000
1	0.0000	0.0000	0.0000	0.0000	0.0000	0.0000	0.0000	0.0000	0.0000

and so

$$\begin{aligned} E[W|\bar{X},S] &= n[(T-\mathrm{LSL})^{-2}E_{X_i<T}[(X_i-T)^2|\bar{X},S] \\ &\quad +(\mathrm{USL}-T)^{-2}E_{X_i>T}[(X_i-T)^2|\bar{X},S]] \\ &= n\left(\frac{S}{c_n}\right)^2\Bigg[\{(T-\mathrm{LSL})^{-2} \\ &\quad -(\mathrm{USL}-T)^{-2}\}g_n\left(\frac{c_n(\bar{X}-T)}{S}\right) \\ &\quad +(\mathrm{USL}-T)^{-2}\left\{\left(\frac{X-T}{S}c_n\right)^2+\frac{1}{n-1}\right\}\Bigg] \end{aligned}$$

$$(3.63)$$

The statistic W is, of course, more easily evaluated than would be the MVUE, $E[W|\bar{X},S]$. Table 3.5, and Fig. 3.2

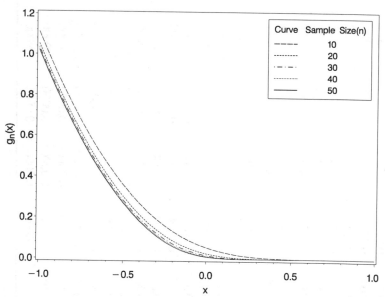

Fig. 3.2 The function $g_n(x)$ for various sample sizes.

giving values of $g_n(z)$, will be of some assistance in this regard. Note that $g_n(0) = \frac{1}{2}(n-1)^{-1}$ and $g_n(-1) = 1 + (n-1)^{-1}$. It can be seen that $g_n(x)$ is small for $x > 0$, while for $x \leqslant -0.25$

$$g_n(x) \cong x^2 + (n-1)^{-1} \tag{3.64}$$

The advantage (reduction of variance) attained by use of the MVUE has not been evaluated as yet. It may be worth the extra effort involved in its calculation and it also provides an example of relevance of 'advanced' statistical methodology to problems connected with quality control. It should be noted that the Blackwell–Rao theorem (see section 1.12.4), on which the derivation of the MVUE is based, is almost 50 years old!

BIBLIOGRAPHY

Boyles, R.A. (1991) The Taguchi capability index, *J. Qual. Technol.*, **23**.

Boyles, R.A. (1992) C_{pm} for asymmetrical tolerances, *Tech. Rep.* Precision Castparts Corp., Portland, Oregon.

Chan, L.K., Cheng, S.W. and Spiring, F.A. (1988a) A new measure of process capability, C_{pm}, *J. Qual. Technol.*, **20**, 160–75.

Chan, L.K., Cheng, S.W. and Spiring, F.A. (1988b) A graphical technique for process capability, *Trans. ASQC Congress*, Dallas, Texas, 268–75.

Chan, L.K., Xiong, Z. and Zhang, D. (1990) On the asymptotic distributions of some process capability indices, *Commun. Statist. – Theor. Meth.*, **19**, 11–18.

Cheng, S.W. (1992) Is the process capable? Tables and graphs in assessing C_{pm}, *Quality Engineering* **4**, 563–76.

Chou, Y.-M., Owen, D.B. and Borrego, S.A. (1990) Lower confidence limits on process capability indices, *Quality Progress*, **23**(7), 231–236.

David, H.A., Hartley, H.O. and Pearson, E.S. (1954) Distribution of the ratio, in a single normal sample, of range to standard deviation *Biometrika*, **41**, 482–93.

Grant, E.L. and Leavenworth, R.S. (1988) *Statistical Quality Control*, (6th edn), McGraw-Hill: New York.

Gupta, A.K. and Kotz, S. (1993) *Tech. Rep.* Bowling Green State University, Bowling Green Ohio.

Haight, F.A. (1967) *Handbook of the Poisson Distribution*, Wiley: New York.

Hsiang, T.C. and Taguchi, G. (1985) *A Tutorial on Quality Control and Assurance – The Taguchi Methods.* Amer. Statist. Assoc. Annual Meeting, Las Vegas, Nevada (188 pp.).

Johnson, N.L., Kotz, S. and Kemp, A.W. (1993) *Distributions in Statistics – Discrete Distributions* (2nd edn), Wiley: New York.

Johnson, T. (1991) A new measure of process capability related to C_{pm}, MS, Dept. Agric. Resource Econ., North Carolina State University, Raleigh, North Carolina.

Kane, V.E. (1986) Process capability indices, *J. Qual. Technol.*, **18**, 41–52.

Kushler, R. and Hurley, P. (1992) Confidence bounds for capability indices, *J. Qual. Technol.*, **24**, 188–195.

Marcucci, M.O. and Beazley, C.F. (1988) Capability indices: Process performance measures, *Trans. ASQC Congress*, 516–23.

Mirabella, J. (1991) Determining which capability index to use, *Quality Progress*, **24**(8), 10.

Patnaik, P.B. (1949) The non-central χ^2- and F-distributions and their applications, *Biometrika*, **36**, 202–32.

Pearn, W.L., Kotz, S. and Johnson, N.L. (1992) Distributional and inferential properties of process capability indices, *J. Qual. Technol.*, **24**, 216–31.

Spiring, F.A. (1989) An application of C_{pm} to the toolwear problem, *Trans. ASQC Congress*, Toronto, 123–8.

Spiring, F.A. (1991a) Assessing process capability in the presence of systematic assignable cause, *J. Qual. Technol.*, **23**, 125–34.

Spiring, F.A. (1991b) The C_{pm} index, *Quality Progress*, **24**(2), 57–61.

Subbaiah, P. and Taam, W. (1991) Inference on the capability index: C_{pm}, MS, Dept. Math. Sci., Oakland University, Rochester, Minnesota.

4

Process capability indices under non-normality: robustness properties

4.1 INTRODUCTION

The effects of non-normality of the distribution of the measured characteristic, X, on properties of PCIs have not been a major research item until quite recently, although some practitioners have been well aware of possible problems in this respect.

In his seminal paper, Kane (1986) devoted only a short paragraph to these problems, in a section dealing with 'drawbacks'.

'A variety of processes result in a non-normal distribution for a characteristic. It is probably reasonable to expect that capability indices are somewhat sensitive to departures from normality. Alas, it is possible to estimate the percentage of parts outside the specification limits, either directly or with a fitted distribution. This percentage can be related to an equivalent capability for a process having a normal distribution.'

A more alarmist and pessimistic assessment of the 'hopelessness' of meaningful interpretation of PCIs (in particular, of C_{pk}) when the process is not normal is presented in Parts 2 and 3 of Gunter's (1989) series of four papers. Gunter emphasizes the difference between 'perfect' (precisely normal)

and 'occasionally erratic' processes, e.g., contaminated processes with probability density functions of form

$$p\varphi(x; \xi_1, \sigma_1) + (1-p)\varphi(x; \xi_2, \sigma_2) \qquad (4.1)$$

where $0 < p < 1; \sigma_1, \sigma_2 > 0; (\xi_1, \sigma_1) \not\equiv (\xi_2, \sigma_2)$ and

$$\varphi(x; \xi, \sigma) = (\sqrt{2\pi}\sigma)^{-1} \exp\left\{ -\frac{1}{2}\left(\frac{x-\xi}{\sigma}\right)^2 \right\}$$

Usually p is close to 1 and $(1-p)$ is small; the second
component in (4.1) represents 'contamination' of the basic
distribution, which corresponds to the first component.

It is often hard to distinguish such types of non-normality, in
practice, by means of standard, conventional testing procedures, though they can cause the behaviour of C_{pk} to vary
quite drastically.

The discussion of non-normality falls into two main parts.
The first, and easier of the two, is investigation of the properties
of PCIs and their estimators when the distribution of X has
specific non-normal forms. The second, and more difficult, is
development of methods of allowing for non-normality and
consideration of use of new PCIs specially designed to be *robust*
(i.e. not too sensitive) to non-normality.

There is a point of view (with which we disagree, in so far as
we can understand it) which would regard our discussion in the
next three sections as being of little importance. McCoy (1991)
regards the manner in which PCIs are used, rather than the
effect of non-normality upon them, as of primary importance,
and writes: 'All that is necessary are statistically generated
control limits to separate residual noise from a statistical signal
indicating something unexpected or undesirable is occurring'.
He also says that PCIs (especially C_{pk}) are dimensionless and
thus '... become(s) a handy tool for simultaneously looking at
all characteristics of conclusions relative to each other'. We are
not clear what this means, but it is not likely that a single index

will provide such an insight into several features of a distribution. Indeed one of us (NLJ) has some sympathy with the viewpoint of M. Johnson (1992), who asserts that '... none of the indices C_p, C_{pl}, C_{pu}, C_{pm} and C_{pk} adds any knowledge or understanding beyond that contained in the equivalent basic parameters μ, σ, target value and the specification limits'.

McCoy (1991, p. 50) also notes: 'In fact, the number of real-life production runs that are normal enough to provide truly accurate estimates of population distribution are more likely than the exceptions', although (p. 51) '... a well-executed study on a stable process will probably result in a normal distribution'.

4.2 EFFECTS OF NON-NORMALITY

Gunter (1989) has studied the interpretation of C_{pk} under three different non-normal distributions. These distributions all have the same mean (ξ) and standard deviation (σ), they therefore all have the same values for C_{pk}, as well as C_p. The three distributions are:

1. a skew distribution with finite lower boundary – namely a $\chi^2_{4.5}$ distribution (Chi-square with 4.5 degrees of freedom)
2. a heavy-tailed ($\beta_2 > 3$) distribution – namely a t_8 distribution
3. a uniform distribution.

In each case, the distribution is standardized (shifted and scaled to produce common values ($\xi = 0$, $\sigma = 1$) for mean and standard deviation).

The proportions (ppm) of NC items outside limits $\pm 3\sigma$ are:(approximately)

For (1.) 14 000 (all above 3σ).
For (2.) 4 000 (half above 3σ; half below -3σ).
For (3.) zero!

And for normal, 2700 (half above 3σ, half below -3σ).
Figure 4.1 shows the four distribution curves.

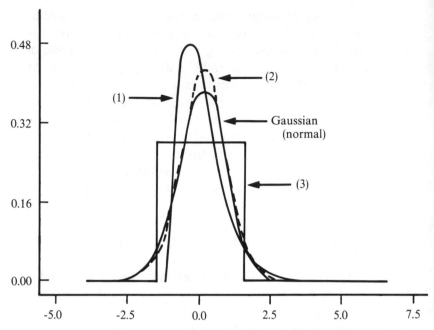

Fig. 4.1 Four different shaped process distributions with the same ξ and σ, and hence the same C_{pk}. (Redrawn from Gunter, B.H. (1989), *Quality Progress*, **22**, 108–109.)

A number of investigations into effects of non-normality on estimators of PCIs have been published. English and Taylor (1990) carried out extensive Monte Carlo (simulation) studies of the distribution of $\hat{C}_p$ (with $C_p = 1$) for normal, symmetrical triangular, uniform and exponential distribution*, with sample sizes $n = 5, 10 (10) 30, 50$. The results are summarized in part in Table 4.1. From these values of $\Pr[\hat{C}_p \geqslant c]$ we note, *inter alia*, that for n less than 20, there can be substantial departures from the true C_p value (as well as differences in interpretations of these values for non-normal distributions).

---

*All four distributions have the same expected value ($\xi = 0$), and standard deviations ($\sigma = 1$). The specification range was LSL $= -3$ to USL $= 3$, so the value of C_p was 1 for all four distributions.

Table 4.1 Values of $\Pr[\hat{C}_p \geqslant c]$ from simulation ($C_p = 1$)

Distribution of X		Normal	Triangular	Uniform	Exponential
$n =$	$c =$				
5	0.50	0.996	1.000	1.000	0.969
	0.75	0.873	0.881	0.924	0.846
	1.00	0.600	0.561	0.522	0.683
	1.25	0.373	0.320	0.267	0.527
	1.50	0.223	0.193	0.152	0.404
	2.00	0.089	0.079	0.058	0.240
	2.50	0.041	0.039	0.025	0.150
10	0.50	1.000	1.000	1.000	0.985
	0.75	0.933	0.959	0.989	0.865
	1.00	0.568	0.529	0.501	0.644
	1.25	0.243	0.189	0.127	0.423
	1.50	0.091	0.064	0.035	0.259
	2.00	0.013	0.011	0.005	0.097
	2.50	0.002	0.003	0.001	0.039
20	0.50	1.000	1.000	1.000	0.996
	0.75	0.979	0.994	1.000	0.900
	1.00	0.548	0.523	0.508	0.617
	1.25	0.129	0.087	0.041	0.315
	1.50	0.017	0.011	0.002	0.135
	2.00	0.001	0.000	0.000	0.022
	2.50	0.000	0.000	0.000	0.003
50	0.50	1.000	1.000	1.000	1.000
	0.75	0.999	1.000	1.000	0.995
	1.00	0.537	0.507	0.497	0.577
	1.25	0.022	0.012	0.002	0.165
	1.50	0.000	0.000	0.000	0.026
	2.00	0.000	0.000	0.000	0.001
	2.50	0.000	0.000	0.000	0.000

Also, the values for the (skew) exponential distributions differ sharply from the values for the three other (symmetrical) distributions especially for larger values of c.

On the more theoretical side, Kocherlakota *et al.* (1992)

have established the distribution of $\hat{C}_p = \frac{1}{3}d/\hat{\sigma}$ in two cases: when the process distribution is (i) contaminated normal with $\sigma_1 = \sigma_2 = \sigma$, see (4.1); and (ii) Edgeworth series

$$f_X(x) = [1 - \tfrac{1}{6}\lambda_3 D^3 + \tfrac{1}{24}\lambda_4 D^4 + \tfrac{1}{72}\lambda_3^2 D^6]\varphi(x;0,1) \qquad (4.2)$$

where D^j signifies jth derivative with respect to X, and

$$\lambda_3 = \frac{\mu_3(X)}{\{\mu_2(X)\}^{\frac{3}{2}}} = \sqrt{\beta_1} \qquad (4.3\,a)$$

and

$$\lambda_4 = \frac{\mu_4(X)}{\{\mu_2(X)\}^2} - 3 = \beta_2 - 3 \qquad (4.3\,b)$$

are standardized measures of skewness and kurtosis respectively. (See section 1.1 for definitions of the central moments $\mu_r(X)$ and the shape factors $\lambda_3, \lambda_4, \sqrt{\beta_1}, \beta_2$.)

4.2.1 Contaminated normal

We will refer in more detail, later, to the results of Kochlerlakota *et al.* (1992). First, we will describe an approach used by ourselves (Kotz and Johnson (1993)) to derive moments of $\hat{C}_p$ for a more general contamination model, with k components (but with each component having the same variance, σ^2), with PDF

$$\sum_{j=1}^{k} p_j \varphi(x; \xi_j, \sigma) \qquad (4.4)$$

The essential clue, in deriving expressions for the moments of $\hat{C}_p$ for this model, is the following simple observation.

A random sample of size n from a population with PDF (4.4) can be regarded as a mixture of random samples of sizes $N_1, \ldots, N_k$ from k populations with PDFs $\varphi(x; \xi_1, \sigma), \ldots, \varphi(x; \xi_k, \sigma)$, where $\mathbf{N} = (N_1, \ldots, N_k)$ have a joint multinomial distribution with parameters $(n; p_1, \ldots, p_k)$ so that

$$\Pr[\mathbf{N} = \mathbf{n}] = \Pr\left[\bigcap_{j=1}^{k} (N_j = n_j)\right]$$

$$= \frac{n!}{\displaystyle\prod_{j=1}^{k} n_j!} \prod_{j=1}^{k} p_j^{n_j}$$

$$\sum_{j=1}^{k} n_j = n \quad \sum_{j=1}^{k} p_j = 1 \quad \mathbf{n} \equiv (n_1, \ldots, n_k)$$

$$(4.5)$$

The conditional distribution of the statistic

$$\sum_{j=1}^{n} (X_j - \xi_0)^2 \tag{4.6}$$

(where ξ_0 is an arbitrary, fixed number), given $\mathbf{N} = \mathbf{n}$, is that of $\sigma^2 \times$ [noncentral chi-squared with n degrees of freedom and noncentrality

parameter $\lambda_n^* = \displaystyle\sum_{j=1}^{k} n_j (\xi_j - \xi_0)^2$] — symbolically, $\sigma^2 \chi_n'^2(\lambda_n^*)$

$$(4.7)$$

The conditional distribution of

$$(n-1)\hat{\sigma}^2 = \sum_{j=1}^{n-1} (X_j - \bar{X})^2$$

is that of

$$\sigma^2 \chi_{n-1}'^2(\lambda_{\mathbf{n}}) \tag{4.8}$$

with

$$\lambda_{\mathbf{n}} = \sum_{j=1}^{k} n_j(\xi_j - \bar{\xi}_{\mathbf{n}})^2 \quad \text{where} \quad \bar{\xi}_{\mathbf{n}} = \frac{1}{n} \sum_{j=1}^{k} n_j \xi_j$$

Expressions for moments of noncentral chi-squared distributions were presented in section 1.5 and have been used in Chapters 2 and 3. We will now use these expressions to derive formulas for the moments of

$$\hat{C}_{\mathrm{p}} = \frac{d}{3\hat{\sigma}}$$

Conditional on $\mathbf{N} = \mathbf{n}$, $\hat{C}_{\mathrm{p}}$ is distributed as

$$\frac{1}{3} \cdot \frac{d(n-1)^{\frac{1}{2}}}{\sigma} / \chi_{n-1}'(\lambda_{\mathbf{n}}) = C_{\mathrm{p}}(n-1)^{\frac{1}{2}} / \chi_{n-1}'(\lambda_{\mathbf{n}}) \tag{4.9}$$

Recalling the formula for moments of noncentral chi-squared from Chapter 1, we have

$$\mathrm{E}[\hat{C}_p^r | \mathbf{N} = \mathbf{n}] = C_p^r (n-1)^{\frac{1}{2}r} \exp(-\tfrac{1}{2}\lambda_{\mathbf{n}})$$

$$\times \sum_{i=0}^{\infty} \left\{ \frac{(\frac{1}{2}\lambda_{\mathbf{n}})^i}{i!} \right\} \frac{\Gamma\left(\dfrac{n-1}{2} + i - \dfrac{r}{2}\right)}{2^{r/2}\Gamma\left(\dfrac{n-1}{2} + i\right)} \tag{4.10}$$

and in particular

$$E[\hat{C}_p|\mathbf{N}=\mathbf{n}]=C_p(n-1)^{\frac{1}{2}}\exp(-\tfrac{1}{2}\lambda_n)$$

$$\times\sum_{i=0}^{\infty}\frac{(\tfrac{1}{2}\lambda_n)^i}{i!}\frac{\Gamma\left(\dfrac{n}{2}+i-1\right)}{\sqrt{2}\Gamma\left(\dfrac{n-1}{2}+i\right)} \qquad (4.11\,a)$$

$$E[\hat{C}_p^2|\mathbf{N}=\mathbf{n}]=C_p^2(n-1)\exp(-\tfrac{1}{2}\lambda_n)$$

$$\times\sum_{i=0}^{\infty}\frac{(\tfrac{1}{2}\lambda_n)^i}{i!}\frac{1}{n-3+2i} \qquad (4.11\,b)$$

Averaging over the distribution of $\mathbf{N}$, we obtain

$$E[\hat{C}_p^r]=\sum_{\mathbf{n}}\left(\frac{n!}{\displaystyle\prod_{j=1}^{k}n_j!}\right)\left(\prod_{j=1}^{k}p_j^{n_j}\right)E[\hat{C}_p^r|\mathbf{n}] \quad r=1,2,\dots \quad (4.12)$$

The summation is over all $n_j\geqslant0$, constrained by

$$\sum_{j=1}^{k}n_j=n.$$

In the special symmetric case of $k=3$, with $\xi_1/\sigma=-a$, $\xi_2=0$, $\xi_3/\sigma=a$, $(a>0)$ and $p_1=p_3=p$, say, the noncentrality parameter is

$$\lambda_n=\left\{n_1+n_3-\frac{1}{n}(n_3-n_1)^2\right\}a^2 \qquad (4.13)$$

Some values of $E[\hat{C}_p]/C_p$ and S.D.$(\hat{C}_p)/C_p$, calculated from (4.11 a) and (4.11 b), are shown in Table 4.2 (based on Kotz

Table 4.2 Normal mixtures expected value (E) and standard deviation (S.D.) of $\hat{C}_p$ when $C_p = 1$

a =	0.005		0.01		0.125		0.25		0.5		1.0	
	E	S.D.	E	S.D.	E	S.D.	E	S.D.	E	S.D.	E	S.D.
a: $k=3$; $\xi_2 - \xi_1 = \xi_3 - \xi_2 = a\sigma$												
n												
$p_1 = p_3 =$												
10 0.05	1.094	0.297	1.094	0.297	1.093	0.297	1.092	0.296	1.081	0.294	1.045	0.286
0.25	1.094	0.297	1.094	0.297	1.090	0.296	1.078	0.293	1.031	0.280	0.890	0.237
20 0.05	1.042	0.180	1.042	0.180	1.041	0.180	1.038	0.179	1.029	0.178	0.994	0.174
0.25	1.042	0.180	1.042	0.180	1.038	0.179	1.026	0.177	0.982	0.169	0.849	0.143
30 0.05	1.028	0.140	1.027	0.140	1.026	0.140	1.024	0.140	1.014	0.193	0.880	0.136
0.25	1.028	0.140	1.027	0.140	1.023	0.140	1.011	0.138	0.968	0.132	0.837	0.112
b: $k=2$; $\lvert \xi_2 - \xi_1 \rvert = a\sigma$												
n												
$p =$												
10 0.1	1.094	0.297	1.094	0.297	1.093	0.297	1.091	0.296	1.082	0.294	1.049	0.287
0.5	1.094	0.297	1.094	0.297	1.092	0.297	1.086	0.295	1.061	0.288	0.976	0.261
20 0.1	1.042	0.180	1.042	0.180	1.041	0.180	1.039	0.179	1.030	0.178	0.998	0.174
0.5	1.042	0.180	1.042	0.180	1.040	0.179	1.034	0.179	1.010	0.174	0.931	0.158
30 0.1	1.027	0.140	1.027	0.140	1.026	0.140	1.024	0.140	1.015	0.139	0.984	0.136
0.5	1.027	0.140	1.027	0.140	1.025	0.140	1.019	0.139	0.996	0.136	0.918	0.123

and Johnson (1992)) for selected values of a and p. (Hung and Hagen (1992) have constructed a computer program using GAUSS for calculation of these quantities.)

The table is subdivided into two parts. In both parts $C_p(=\frac{1}{3}d/\sigma)=1$. In Table 4.2$a$ the cases $k=3$, $n=n_1+n_2+n_3=10, 20$, and 30; $p_1=p_3=0.05$, and 0.25; $a=0.005, 0.01, 0.125, 0.25, 0.5$, and 1.0 are presented. Table 4.2b covers situations when $k=2$ for the same values of n and a, with $p=0.1$, and 0.5.

From the tables we note that

1. for given n, the bias and S.D. both decrease as a increases;
2. for given a, the bias and S.D. both decrease as n increases;
3. for given a and n, the bias and S.D. do not vary greatly with p_1 and p_2 (when $k=3$) or with p (when $k=2$). The bias is numerically greater when the 'contaminating' parameters are smaller;
4. on the other hand, the greater the value of a, the more marked is the variation (such as it is) with respect to p_1 and p_3 (or p).

Observe, also, that the bias is positive when a is small, and decreases as the proportion of contamination increases. However, the bias is negative for larger values of a and p, and it becomes more pronounced as n increases. This is contrary to the situation when $p=0$ (no contamination) in which the bias of $\hat{C}_p$ is always positive, though decreasing as n increases. Gunter (1989) also observed a negative bias, when the contaminating variable has a larger variance than the main one.

Kocherlakota *et al.* (1992) provide more detailed tables for the case $k=2$. They also derived the moments of $\hat{C}_{pu}=(\text{USL}-\bar{X})/(3\hat{\sigma})$. The distribution of $\hat{C}_{pu}$ is essentially a mixture of doubly noncentral t distributions; see section 1.7.

4.2.2 Edgeworth series distributions

The use of Edgeworth expansions to represent mild deviations from normality has been quite fashionable in recent years. There is a voluminous literature on this topic, including Subrahmaniam* (1966, 1968 a, b) who has been a pioneer in this field.

It has to be kept in mind that there are quite severe limitations on the allowable values of $\sqrt{\beta_1}$ and β_2 to ensure that the PDF (4.2) remains positive for all x; see, e.g. Johnson and Kotz (1970, p. 18), and for a more detailed analysis, Balitskaya and Zolotuhina (1988)). Kocherlakota *et al.* (1992) show that for the process distribution (4.2) the expected value of $\hat{C}_p$ is

$$E[\hat{C}_p] = \frac{(n-1)^{\frac{1}{2}}\Gamma(\frac{1}{2}(n-2))}{\sqrt{2}\,\Gamma(\frac{1}{2}(n-1))} C_p \{1 + \tfrac{3}{8}g_n(\beta_2 - 3) - \tfrac{3}{8}h_n\beta_1\}$$

$$(4.14\,a)$$

where

$$g_n = \frac{n-1}{n(n+1)}$$

and

$$h_n = \frac{(n-2)}{n(n+1)(n+3)}$$

The premultiplier of C_p will be recognized as the bias correlation factor b_{n-1}, defined in (2.16), so (4.14 a) can be written

$$E[\hat{C}_p] = E[\hat{C}_p|\text{normal}]\{1 + \tfrac{3}{8}g_n(\beta_2 - 3) - \tfrac{3}{8}h_n\beta_1\} \quad (4.14\,b)$$

---

*K. Subrahmaniam was the (pen) name used by K. Kocherlakota in the 1960s.

Kocherlakota *et al.* (1992) also show that the variance of $\hat{C}_p$ is

$$\frac{n-1}{n(n-3)} C_p^2 \left[1 + \frac{1}{n}\{1 + g_n(\beta_2 - 3) - \tfrac{1}{9}h_n\beta_1\} \right] - \{E[\hat{C}_p]\}^2 \tag{4.14 c}$$

Table 4.3 is based on Table 8 of Kocherlakota *et al.* (1992).

As is to be expected from section 1.1, for given β_2 the values of $E[\hat{C}_p]/C_p$ do not vary with $\sqrt{\beta_1}$, and the values of S.D.$(\hat{C}_p)/C_p$ vary only slightly with $\sqrt{\beta_1}$. Also, for given $\sqrt{\beta_1}$, the values of S.D.$(\hat{C}_p)/C_p$ increase with β_2.

Kocherlakota *et al.* (1992) carried out a similar investigation for the distribution of the natural estimator of C_{pu} $(=\tfrac{1}{3}(\text{USL} - \xi)/\sigma)$,

$$\hat{C}_{pu} = \frac{\text{USL} - \bar{X}}{3S} \tag{4.15}$$

Table 4.3 Expected value and standard deviation of $\hat{C}_p/C_p$ for Edgeworth process distributions

$n=$	$\sqrt{\beta_1}$	β_2	$E[\hat{C}_p]/C_p$	S.D. $(\hat{C}_p)/C_p$
10	0.0	3.0	1.094	0.297
		4.0	1.128	0.347
		5.0	1.161	0.389
	± 0.4	3.0	1.094	0.300
		4.0	1.128	0.349
		5.0	1.161	0.392
30	0.0	3.0	1.027	0.140
		4.0	1.039	0.169
		5.0	1.051	0.193
	± 0.4	3.0	1.027	0.141
		4.0	1.039	0.170
		5.0	1.051	0.194

If the process is symmetrical (so that $\beta_1 = 0$), $E[\hat{C}_{pu}]$ is proportional to C_{pu}. Table 4.4 (based on Kocherlakota *et al.*

Table 4.4 Expected value and standard deviation of $\hat{C}_{pu}$ for Edgeworth process distributions

n	$\sqrt{\beta_1}$	β_2	$E[\hat{C}_{pu}]$ $C_{pu}=1.0$	$E[\hat{C}_{pu}]$ $C_{pu}=1.5$	$S.D.(\hat{C}_{pu})$ $C_{pu}=1.0$	$S.D.(\hat{C}_{pu})$ $C_{pu}=1.5$
10	−0.4	3.0	1.086	1.632	0.287	0.426
		4.0	1.119	1.682	0.337	0.502
		5.0	1.153	1.733	0.381	0.569
	0.0	3.0	1.094	1.641	0.320	0.433
		4.0	1.128	1.592	0.366	0.533
		5.0	1.161	1.742	0.407	0.596
	0.4	3.0	1.100	1.647	0.345	0.486
		4.0	1.134	1.697	0.388	0.554
		5.0	1.107	1.748	0.427	0.615
30	−0.4	3.0	1.024	1.538	0.136	0.201
		4.0	1.036	1.556	0.166	0.246
		5.0	1.048	1.574	0.190	0.283
	0.0	3.0	1.027	1.540	0.154	0.220
		4.0	1.039	1.558	0.180	0.261
		5.0	1.051	1.576	0.203	0.297
	0.4	3.0	1.029	1.542	0.168	0.235
		4.0	1.041	1.560	0.193	0.274
		5.0	1.053	1.578	0.214	0.308

(1992)) shows that even when β_1 is not zero, $E[\hat{C}_{pu}]$ is very nearly proportional to C_{pu}. Of course, if β_1 is large, it will not be possible to represent the distribution in the Edgeworth form (4.2). Table 4.3 exhibits reassuring evidence of robustness of $\hat{C}_p$ to mild skewness.

4.2.3 Miscellaneous

Price and Price (1992) present values, estimated by simulation, of the expected values of $\hat{C}_p$ and $\hat{C}_{pk}$, from the following process distributions, numbered [1] to [13]

Distribution number and name		Skewness ($\sqrt{\beta_1}$)
[1]	Normal (50, 1)	0
[2]	Uniform (48.268, 51.732)	0
[3]	10 × Beta (4.4375, 13.3125) + 47.5	0.506
[4]	10 × Beta (13.3125, 4.4375) + 42.5	−0.506
[5]	Gamma (9, 3) + 47	0.667
[6]	Gamma (4, 2) + 48	1.000
[7]	($\chi^2_{4.5}$) Gamma (2.25, 1.5) + 48.5	1.333
[8]	(Exponential) Gamma (1, 1) + 49	2.000
[9]	Gamma (0.75, 0.867) + 49.1340	2.309
[10]	Gamma (0.5, 0.707) + 49.2929	2.828
[11]	Gamma (0.4, 0.6325) + 49.3675	3.163
[12]	Gamma (0.3, 0.5477) + 49.4523	3.651
[13]	Gamma (0.25, 0.5) + 49.5	4.000

In all cases; the expected value of the process distribution is 50, and the standard deviation is 1.

The values shown in the tables below are based on computer output kindly provided by Drs Barbara and Kelly Price of Wayne State University in 1992.

The specification limits are shown in these tables, together with the appropriate values of C_p and C_{pk}. The symbols (M), (L) and (R) indicate that $T =, >, < \frac{1}{2}(LSL + USL)$ respectively. In all cases, $T = 50$.

	LSL	USL	C_p	C_{pk}		LSL	USL	C_p	C_{pk}
(M)	48.5	51.5	0.5	0.5	(L)	44.0	53.0	1.0	1.5
(L)	45.5	51.5	0.5	1.0	(R)	47.0	56.0	1.0	1.5
(R)	48.5	54.5	0.5	1.0	(M)	44.0	56.0	2.0	2.0
(M)	47.0	53.0	1.0	1.0	(L)	41.0	56.0	2.0	2.5
					(R)	44.0	59.0	2.0	2.5

The value of $E[\hat{C}_p/C_p]$ does not depend on ξ, so we do not need to distinguish (M), (L) and (R), nor does it depend on C_p. Hence the simple Table 4.5 suffices. In this table, the distributions are arranged in increasing order of $|\sqrt{\beta_1}|$.

Table 4.5 Simulated values of $E[\hat{C}_p/C_p]$

Distribution	$n=10$	$n=30$	$n=100$
[1] (normal)	1.1183	1.0318	1.0128
[2] (Uniform)	1.0420	1.0070	1.0017
[3] (Beta) ⎱	1.1171	1.0377	1.0137
[4] (Beta) ⎰			
[5] (Gamma)	1.1044	1.0297	1.0091
[6] (Gamma)	1.1155	1.0371	1.0143
[7] (Gamma)	1.1527	1.0474	1.0146
[8] (Gamma)	1.2714	1.0801	1.0242
[9] (Gamma)	1.3478	1.1155	1.0449
[10] (Gamma)	1.5795	1.1715	1.0449
[11] (Gamma)	1.6220	1.2051	1.0664
[12] (Gamma)	1.8792	1.2595	1.0850
[13] (Gamma)	2.2152	1.2869	1.0966

These values do not depend on T. The values for the two beta distributions have been combined, as they should be the same.

Comparing the estimates for the normal distribution ([1]) with the correct values $(1.0942, 1.0268, 1.0077$ for $n=10, 30, 100$ respectively) we see that the estimates are in excess by about 2% for $n=10$, $\frac{1}{2}$% for $n=30$ or 100.

Sampling variation is also evident in the progression of values for the nine gamma distributions, especially in the $n=100$ values for distributions [6] and [7], and [9] and [10]. As skewness increases, so does $E[\hat{C}_p/C_p]$, reaching quite remarkable values for high values of $\sqrt{\beta_1}$. These however, correspond to quite remarkable (and, we hope and surmise, rare) forms of process distributions, having even higher values of $\sqrt{\beta_1}$ than does the exponential distribution. For moderate skewness, the values of $E[\hat{C}_p/C_p]$ are quite close to those of the normal distribution (c.f. Table 4.3).

Table 4.6 presents values of $E[\hat{C}_{pk}]$ estimated from simulation for a number of cases, selected from those presented by Price and Price (1992).

We again note the extremely high positive bias – this time

of $\hat{C}_{pk}$ as an estimator of C_{pk} – for distribution [13] when $n = 10$, and only relatively smaller biases when $n = 30$ and $n = 100$.

For the exponential distribution [8] there is a quite substantial positive bias in $\hat{C}_{pk}$. The bias is larger when ξ is greater than $\frac{1}{2}(LSL + USL)$ – case (L) – than when it is smaller – case (R). The greater among these biases for exponential, in Table 4.6, are of order 25–35% (when $n = 10$), falling to $2\frac{1}{2}$–5% when $n = 100$.

As for $\hat{C}_{p}$, the results for the extreme distribution [13] are sharply discrepant from those for normal, and mildly skew distributions. Of course, [13] is included only to exhibit the possibility of such remarkable biases, not to imply that they are anything like everyday occurrences.

Coming to the variability of the estimators $\hat{C}_{p}$ and $\hat{C}_{pk}$, we note that the standard deviation of $\hat{C}_{p}$ might be expected to be approximately proportional to $\sqrt{(\beta_2 - 1)}$ where $\beta_2 (= \mu_4/\sigma^4)$ is the kurtosis shape factor (see section 1.1) for the process distribution. We would therefore expect lower standard deviations for uniform process distributions ($\beta_2 = 1.8$) than for normal ($\beta_2 = 3$) and higher standard deviations when $\beta_2 > 3$ (e.g. for the exponential [8] with $\beta_2 = 9$). Values of $\sqrt{(\beta_2 - 1)}$ are included in Table 4.7, and the estimated standard deviations support these conjectures.

It should be realized that Tables 4.5–7, being based on simulations, can give only a broad global picture of variation in bias and standard deviation of $\hat{C}_{p}$ and $\hat{C}_{pk}$ with changes in the shape of the process distribution. More precise values await completion of relevant mathematical analyses which we hope interested readers might undertake.

4.3 CONSTRUCTION OF ROBUST PCIs

The PCIs described in this section are not completely distribution-free, but are intended to reduce the effects of non-normality.

Table 4.6 Values of $E[\hat{C}_{pk}]$ from simulation (Price and Price (1992))

C_{pk}	$\dfrac{\min(\xi - LSL, USL - \xi)}{d}$	Process distribution*	$n = 10$	$n = 30$	$n = 100$
0.5	1	[1]	0.464	0.468	0.480
		[2]	0.424	0.453	0.474
		[3][4]	0.464	0.471	0.481
		[7]	0.480	0.474	0.480
		[8]	0.527	0.487	0.485
		[13]	0.904	0.582	0.519
	1/2	[1]LR	0.559	0.516	0.506
		[2]LR	0.521	0.504	0.501
		[3]L[4]R/[3]R[4]L	0.569/0.548	0.525/0.509	0.513/0.504
		[7]L/[7]R	0.597/0.555	0.528/0.519	0.510/0.504
		[8]L/[8]R	0.671/0.600	0.548/0.532	0.515/0.509
		[13]L/[13]R	1.259/0.956	0.670/0.617	0.557/0.540
1.0	1	[1]	1.023	0.985	0.986
		[2]	0.945	0.957	0.975
		[3][4]	1.022	0.989	0.988
		[7]	1.057	0.998	0.987
		[8]	1.163	1.028	0.997
		[13]	2.012	1.226	1.067

2/3	[1]LR	1.118	1.032	1.013
	[2]LR	1.042	1.007	1.002
	[3]L[4]R/[3]R[4]L	1.129/1.111	1.043/1.032	1.017/1.011
	[7]L/[7]R	1.174/1.132	1.052/1.043	1.017/1.012
	[8]L/[8]R	1.307/1.236	1.088/1.072	1.027/1.021
	[13]L/[13]R	2.337/2.064	1.313/1.261	1.105/1.088
2.0	[1]	2.141	2.016	1.999
	[2]	1.987	1.964	1.976
	[3][4]	2.139	2.027	1.998
	[7]	2.209	2.045	2.002
	[8]	2.435	2.108	2.021
	[13]	4.227	2.513	2.164
4/5	[1]LR	2.237	2.064	2.026
	[2]LR	2.084	2.014	2.003
	[3]L[4]R/[3]R[4]L	2.245/2.233	2.081/2.070	2.030/2.024
	[7]L/[7]R	2.326/2.284	2.099/2.090	2.032/2.026
	[8]L/[8]R	2.578/2.507	2.168/2.153	2.052/2.046
	[13]L/[13]R	4.582/4.279	2.600/2.548	2.201/2.185

*Notes:

(a) When $\min(\xi - \mathrm{LSL}, \mathrm{USL} - \xi)/d = 1$, we have $\xi = \frac{1}{2}(\mathrm{LSL} + \mathrm{USL})$, so only (M) applies.

(b) Since [3] and [4] are mirror images, results can be merged as shown.

(c) For symmetrical distributions (normal [1] and uniform [2]), the L and R values should be the same and so they have been averaged.

Table 4.7 Estimates of S.D. $(\hat{C}_p)$ and S.D. $(\hat{C}_{pk})$ from simulation (Price and Price (1992))

Distribution	$(\beta_2 - 1)^{\frac{1}{2}}$	sgn $\{\xi - \frac{1}{2}(LSL + USL)\}$	n = 30		n = 100	
			S.D.$(\hat{C}_p/C_p)$	S.D.$(\hat{C}_{pk})$	S.D.$(\hat{C}_p/C_p)$	S.D.$(\hat{C}_{pk})$
Normal [1]	1.414	0(M) 1(L), −1(R)	0.148	$\left\{\begin{array}{l}0.148\\0.160\end{array}\right.$	0.075	$\left\{\begin{array}{l}0.077\\0.082\end{array}\right.$
Uniform [2]	0.894	0(M) 1(L), −1(R)	0.089	$\left\{\begin{array}{l}0.088\\0.110\end{array}\right.$	0.045	$\left\{\begin{array}{l}0.047\\0.056\end{array}\right.$
$\chi^2_{4.5}$ [7]	2.160	0(M) 1(L) −1(R)	0.199	$\left\{\begin{array}{l}0.192\\0.245\\0.166\end{array}\right.$	0.111	$\left\{\begin{array}{l}0.109\\0.137\\0.092\end{array}\right.$
Exponential [8]	2.828	0(M) 1(L) −1(R)	0.273	$\left\{\begin{array}{l}0.258\\0.237\\0.255\end{array}\right.$	0.142	$\left\{\begin{array}{l}0.139\\0.169\\0.119\end{array}\right.$

Values for S.D.$(\hat{C}_{pk})$ correspond to process distributions with $C_{pk} = 1$.

$$\text{sgn}(h) = \begin{cases} -1 & \text{for} \quad h < 0 \\ 0 & \text{for} \quad h = 0 \\ +1 & \text{for} \quad h > 0 \end{cases}$$

Table 4.8 0.135% and 99.865% points of standardized Pearson curves with positive skewness ($\sqrt{\beta_1}>0$). If $\sqrt{\beta_1}<0$, interchange 0.135% and 99.865% points and reverse signs. (Clements (1989))

β_2 \\ $\sqrt{\beta_1}$	0.0	0.2	0.4	0.6	0.8	1.0	1.2	1.4	1.6	1.8
1.8	−1.727	−1.496	−1.230	−0.975	−0.747	—	—	—	—	—
	1.727	1.871	1.896	1.803	1.636	—	—	—	—	—
2.2	−2.210	−1.912	−1.555	−1.212	−0.927	−0.692	—	—	—	—
	2.210	2.400	2.454	2.349	2.108	1.822	—	—	—	—
2.6	−3.000	−2.535	−1.930	−1.496	−1.125	−0.841	−0.616	—	—	—
	3.000	2.869	2.969	2.926	2.699	2.314	1.928	—	—	—
3.0	−3.000	−2.689	−2.289	−1.817	−1.356	−1.000	−0.739	−0.531	—	—
	3.000	3.224	3.358	3.385	3.259	2.914	2.405	1.960	—	—
3.4	−3.261	−2.952	−2.589	−2.127	−1.619	−1.178	−0.865	−0.634	—	—
	3.261	3.484	3.639	3.675	3.681	3.468	2.993	2.398	—	—
3.8	−3.458	−3.118	−2.821	−2.396	−1.887	−1.381	−1.000	−0.736	−0.533	—
	3.458	3.678	3.844	3.951	3.981	3.883	3.861	2.945	2.322	—
4.2	−3.611	−3.218	−2.983	−2.616	−2.132	−1.602	−1.149	−0.840	−0.617	—
	3.611	3.724	3.997	4.124	4.194	4.177	3.496	3.529	2.798	—
4.6	−3.731	−3.282	−3.092	−2.787	−2.345	−1.821	−1.316	−0.950	−0.701	−0.510
	3.731	3.942	4.115	4.253	4.351	4.386	4.311	4.015	3.364	2.609
5.0	−3.828	−3.325	−3.167	−2.914	−2.524	−2.023	−1.494	−1.068	−0.785	−0.580
	3.828	4.034	4.208	4.354	4.468	4.539	4.532	4.372	3.907	3.095

For each ($\sqrt{\beta_1}$, β_2) combination, the upper row contains 0.135% points (θ_l) and the lower, 99.865% points (θ_u).

4.3.1 Clements' method

Clements (1989) proposed a method of construction based on the assumption that the process distribution can adequately be represented by a Pearson distribution (see section 1.11). Essentially, the aim is to replace the multiplier '6' in the denominator of C_p by a number, θ say, that will be such that

$$\Pr[\xi - \tfrac{1}{2}\theta\sigma \leqslant X \leqslant \xi + \tfrac{1}{2}\theta\sigma] = 0.0027$$

For given values of the skewness and kurtosis coefficients, $\sqrt{\beta_1}$ and β_2, Table 4.8, gives values θ_l, θ_u such that

$$\Pr[X \leqslant \xi - \theta_l\sigma] = 0.135\% = \Pr[X \geqslant \xi + \theta_u\sigma]$$

We then take $\theta = \theta_u - \theta_l$. (These tables were given by Clements, adapting tables in Gruska *et al.* (1989).)

For example, if $\sqrt{\beta_1} = 1.0$ and $\beta_2 = 5.0$, we find $\theta_l = -2.023$ and $\theta_u = 4.539$, so $\theta = 2.023 + 4.539 = 6.572$. If $\sqrt{\beta_1} = -1.0$, $\beta_2 = 5.0$, so we have $\theta_l = -4.539$ and $\theta_u = 2.023$, and so, again, $\theta = 6.572$.)

The index C_p would be calculated as $d/(3.286\sigma)$.

For normal distributions we have $\theta = 6$, leading to the familiar $C_p = \tfrac{1}{3}d/\sigma$.

In calculating C_{pk}, Clements suggests using the same value of θ, though replacing $\bar{X}$ by the median, M, but this is not a really essential feature of his method. Further discussion of Clements' method is contained in Pearn and Kotz (1992).

4.3.2 Johnson–Kotz–Pearn method

Application of Clements' method requires knowledge of the coefficients $\sqrt{\beta_1}$ and β_2, which may not be easily obtainable. Rather large samples are needed for accurate estimation of these quantities. Alternative approaches seek to avoid this difficulty.

Johnson *et al.* (1992) suggest using

$$C_{p(\theta)} = \frac{\text{USL} - \text{LSL}}{\theta\sigma} = \frac{d}{\frac{1}{2}\theta\sigma} \qquad (4.16)$$

where θ is now chosen so that the 'capability' – namely the proportion of conforming items, with optimum choice of ξ – is not greatly affected by the shape of the process distribution. (The original C_p is, in the notation of (4.16), $C_{p(6)}$.)

The results of Pearson and Tukey (1965) are relevant to the choice of θ. Table 4.9 (their Table 7) gives values of θ such that the optimum proportion conforming has specified values, for several chi-squared (Type III) distributions.

One can see that $\theta = 5.15$ will give very stable values for the proportion ($\sim 1\%$) of NC product.

The remarkable stability of θ, indicated in Table 4.6, was discovered (for χ^2 and $\log \chi^2$ distributions) during World War II by the British statistician C.P. Winsor, who communicated his discovery to E.S. Pearson.

In the case of non-normal Pearson distributions (section 1.11), the value $\theta = 4.34$ gives $P = 0.97$ for $0 \leqslant \beta_1 \leqslant 3.111$, $1 \leqslant \beta_2 \leqslant 5.997$.

We recommend use of

$$C_{p(5.15)} = \frac{2d}{5.15\sigma} = \frac{d}{2.575\sigma} \qquad (4.17)$$

It is true that robustness is attained at the cost of abandoning the 0.27% level for expected proportion of NC product, but we feel this is a reasonable exchange. Of course, those who feel that proportion NC is not a major concern should also feel less qualms about variation from the 0.27% standard.

We also note the corresponding variant of C_{pk};

$$C_{pk(5.15)} = \frac{d - |\xi - \frac{1}{2}(\text{LSL} + \text{USL})|}{2.575\sigma} \qquad (4.18)$$

Table 4.9 Chi-squared: values of θ such that $\Pr[\xi - \tfrac{1}{2}\theta\sigma \leqslant X \leqslant \xi + \tfrac{1}{2}\theta\sigma] = P$ for suitable ξ

Degrees of freedom	6	8	10	12	15	20	30	60	∞(normal)
β_1	1.33	1.00	0.80	0.67	0.53	0.40	0.27	0.13	0.00
β_2	5.00	4.50	4.20	4.00	3.80	3.60	3.40	3.20	3.00
θ for $P=0.95$	3.82	3.84	3.85	3.87	3.88	3.89	3.90	3.91	3.92
θ for $P=0.98$	4.60	4.61	4.62	4.62	4.63	4.63	4.64	4.64	4.65
θ for $P=0.99$	5.16	5.15	5.15	5.15	5.15	5.15	5.15	5.15	5.15

Moment estimators for (4.17) and (4.18) are obtained by replacing σ by an appropriate estimator.

At this point we reiterate the difference between Clements' method and the Johnson–Kotz–Pearn method. In the first method there is an attempt to make a direct allowance for the values of the skewness and kurtosis coefficients, while the second method aims at getting limits which are insensitive to these values. In the second method we no longer have guaranteed equal tail probabilities, but we do not have to estimate $\sqrt{\beta_1}$ and β_2 which it may be difficult to achieve with accuracy, because of use of third and fourth sample moments, which are subject to large fluctuations. Both methods rely on the assumption that the population distribution has a unimodal shape close to a Pearson distribution for Clements' method, and more restrictively, close to a gamma distribution for Johnson–Kotz–Pearn method.

Another approach, also aimed at enhancing connection between PCI values and expected proportions NC, tries to 'correct' the PCI, so that the corrected value corresponds (at least approximately) to what would be the value for a normal process distribution with the same expected proportion NC. Munechika (1986, 1992) utilized an approximate relation between corresponding percentiles of standardized normal and Gram-Charlier (Edgeworth) distribution (see Johnson and Kotz (1970, p. 34)) to obtain an approximate relationship between the process PCI and the corrected (equivalent normal) PCI values. He applied this only to the case where there is only an upper specification limit (no LSL), with the C_{pku} index (see equation (2.25b)). The approximation is quite good for gamma process distributions which are not too skew.

From the relationship (in an obvious notation) $X_\alpha \cong U_\alpha + \frac{1}{6}(U_\alpha^2 - 1)\sqrt{\beta_1}$ he obtained 3(process index) $\cong$ 3(corrected index) $+ \frac{1}{6}\{(\text{corrected index})^2 - 1\}$ leading to (corrected index $\cong (1/3\sqrt{\beta_1})[\{\beta_1 + 18(\text{process index})\sqrt{\beta_1} + 9\}^{1/2} - 3]$.

The inverse (also approximate) relationship $U_\alpha \cong X_\alpha - \frac{1}{6}(X_\alpha^2 - 1)\sqrt{\beta_1}$ would give the somewhat simpler formula

(corrected index)$\cong$(process index)$-\frac{1}{18}\{9$(process index)$^2-1\}$
(This formula was not used in Munechika (1986, 1992).)

Of course, use of this correction requires a value for $\sqrt{\beta_1}$. Unless this is well established, it is necessary to estimate it. Such estimation may well be subject to quite large sampling variability.

4.3.3 'Distribution-free' PCIs

Chan *et al.* (1988) proposed the following method of obtaining 'distribution-free' PCIs. They note that the estimator, $\hat{\sigma}$, in the denominator of $\hat{C}_p$ is used solely to estimate the length (6σ) of the interval covering 99.73% of the values of X (on the twin assumptions of normality and perfect centring, at $\xi=\frac{1}{2}$(LSL + USL)). They propose using distribution-free tolerance intervals, as defined, for example, in Guenther (1985) (not to be confused with Gunter (1989)!) to estimate this length. These tolerance intervals (widely used in statistical methodology for the last 50 years) are designed to include at least $100\beta\%$ of a distribution with preassigned probability $100(1-\alpha)\%$, for given β (usually close to 1) and α (usually close to zero). In the PCI framework, a natural choice for β would be $\beta=0.9973$, with, perhaps, $\alpha=0.05$. Unfortunately construction of such intervals would require prohibitively large samples (of size 1000 or more).

Chan *et al.* (1988) suggest that this difficulty can be overcome in the following way. Construction of tolerance intervals with smaller values of β (but still with $\alpha=0.05$) requires smaller samples (less than 300). They recommend taking

- $\beta=0.9546$ and using $\frac{3}{2}\times$(length of tolerance interval) in place of $6\hat{\sigma}$, or
- $\beta=0.6826$ and using $3\times$(length of tolerance interval) in place of $6\hat{\sigma}$.

The bases for their choices are that for a normal distribution

- the interval $(\xi - 2\sigma, \xi + 2\sigma)$ of length 4σ contains 95.46% of values, and
- the interval $(\xi - \sigma, \xi + \sigma)$ of length 2σ contains 68.26% of values (and, of course, $6\sigma = \frac{3}{2} \times 4\sigma = 3 \times 2\sigma$).

It is here that we must part company with them, as it seems unreasonable to use relationships based on normal distributions to estimate values which are supposed to be distribution-free!

4.3.4 Bootstrap methods

Franklin and Wasserman (1991), together with Price and Price (1992) should be regarded as the pioneers of application of bootstrap methodology in estimation of C_{pk}. The bootstrap method was introduced some twelve years ago (see Efron, 1982) and has achieved remarkably rapid acceptance among statistical practitioners since then. (Over 600 papers on the bootstrap method were published in the period 1979–90!) It is not until very recently, however, that its application in the field of PCIs has been developed.

The bootstrap method is a technique whereby an estimate of the distribution function of a statistic based on a sample size n, say, is obtained from data in a random sample, of size m ($\geq n$) say, by 're-sampling' samples of size n – with replacement – from these m values and calculating the corresponding values of the statistic in question. Usually $m = n$, but this need not be the case. Here is a brief formal description of the method.

Given a sample of size m with sample values $x_1, x_2, \ldots, x_m$ we choose (with replacement) a random sample ([1], say) $x_{[1]1}, \ldots, x_{[1]n}$ of size n, and calculate $\hat{C}_{[1]pk}$ from this new 'sample'. This is repeated many (g) times and we obtain a set of values $\hat{C}_{[1]pk}, \hat{C}_{[2]pk}, \ldots, \hat{C}_{[g]pk}$, which we regard as approximating the distribution of $\hat{C}_{pk}$ in samples of size n – this estimate is the bootstrap distribution. (The theoretical basis of this method is that we use the empirical cumulative distribution from the first sample – assigning probability m^{-1} to each value – as an approximation to the true CDF of X.) Practice has indicated that a minimum of 1000 bootstrap samples are needed for a reliable calculation of bootstrap confidence intervals for C_{pk}.

According to Hall (1992) – a leading expert on bootstrap methodology – difficulties 'in applying the bootstrap to process capability indices is that these indices are ratios of random variables, with a significant amount of variability in the variable in the denominator'. (Similar situations exist in regard to estimation of a correlation coefficient, and of the ratio of two expected values. The bootstrap performs quite poorly in these two (better-known) contexts.)

Franklin and Wasserman (1991) distinguish three types of bootstrap confidence intervals for C_{pk}, discussed below.

4.3.4a The 'standard' confidence interval

One thousand bootstrap samples are obtained by re-sampling from the observed values $X_1, X_2, ..., X_n$ (with replacement). The arithmetic mean $\hat{C}^*_{pk}$ and standard deviation $S^*(\hat{C}_{pk})$ of the $\hat{C}_{pk}$s are calculated. The $100(1-\alpha)\%$ confidence interval for C_{pk} is then

$$(\hat{C}^*_{pk} - z_{1-\alpha/2} S^*(\hat{C}_{pk}), \hat{C}^*_{pk} + z_{1-\alpha/2} S^*(\hat{C}_{pk})) \qquad (4.19)$$

where $\Phi(z_{1-\alpha/2}) = 1 - \alpha/2$.

4.3.4b The percentile confidence interval

The 1000 $\hat{C}_{pk}$s are ordered as $\hat{C}_{pk}(1) \leqslant \hat{C}_{pk}(2) \leqslant \cdots \leqslant \hat{C}_{pk}(1000)$. The confidence interval is then $(\hat{C}_{pk}(\langle 500\alpha \rangle), \hat{C}_{pk}(\langle 500(1-\alpha) \rangle))$, where $\langle \rangle$ denotes 'nearest integer to'.

4.3.4c *The bias-corrected percentile confidence interval*

This is intended to produce a shorter confidence interval by allowing for the skewness of the distribution of $\hat{C}_{pk}$. Guenther (1985) pointed out the possibility of doing this in the general case, and Efron (1982) developed a method applicable in bootstrapping situations.

The first step is to locate the observed $\hat{C}_{pk}$ in the bootstrap order statistics $\hat{C}_{pk}(1) \leqslant \cdots \leqslant \hat{C}_{pk}(1000)$. For example, if we have $\hat{C}_{pk} = 1.42$ from the original data, and among the bootstrapped values we find

$$\hat{C}_{pk}(365) = 1.41 \quad \text{and} \quad \hat{C}_{pk}(366) = 1.43$$

then we estimate

$$\Pr[\hat{C}_{pk} \leqslant 1.41] \quad \text{as} \quad \frac{365}{1,000} = 0.365 = p_0, \text{ say}$$

We then calculate $\Phi^{-1}(p_0) = z_0$ (i.e. $\Phi(z_0) = p_0$). In our example, $z_0 = \Phi^{-1}(0.365) = -0.345$.

We next calculate

$$p_l = \Phi(2z_0 - z_{1-\alpha/2}) \quad \text{and} \quad p_u = \Phi(2z_0 + z_{1-\alpha/2})$$

(in our case, with $\alpha = 0.05$; $z_{1-\alpha/2} = \Phi^{-1}(0.975) = 1.960$, $p_l = \Phi(-2.650) = 0.004$; $p_u = \Phi(1.270) = 0.898$, and form the confidence interval

$$(\hat{C}_{pk}(1000p_l), \hat{C}_{pk}(1000p_u)).$$

In our example, this is $(\hat{C}_{pk}(4), \hat{C}_{pk}(898))$.

The rationale of this method is that since the observed $\hat{C}_{pk}$ is not at the median of the bootstrap distribution (in our case, below the median) the confidence limits should be adjusted approximately (in our case, lowered, because $z_0 < 0$). Evidently, the method can be applied to other PCIs.

Schenkler (1985) has presented results casting doubt on the efficiency of the 'percentile' and 'bias-percentile' methods above. Hall (1992) suggests that the method of bootstrap iteration described by Hall and Martin (1988) might be more useful in this case. Hall, *et al.* (1989) describe an application of this method to estimation of correlation coefficients, but, to the best of our knowledge, it has not as yet been tested for estimation of PCIs.

4.4 FLEXIBLE PCIs

Johnson *et al.* (1992) have introduced a 'flexible' PCI, taking into account possible differences in variability above and below the target value, T. They define one-sided PCIs

$$CU_{jkp} = \frac{1}{3\sqrt{2}} \frac{USL - T}{\{E_{X > T}[(X - T)^2]\}^{\frac{1}{2}}} \qquad (4.20\,a)$$

and

$$CL_{jkp} = \frac{1}{3\sqrt{2}} \frac{T - LSL}{\{E_{X < T}[(X - T)^2]\}^{\frac{1}{2}}} \qquad (4.20\,b)$$

where (as in section 3.5)

$$E_{X > T}[(X - T)^2] = E[(X - T)^2 | X > T]\Pr[X > T] \quad (4.21\,a)$$

and

$$E_{X < T}[(X - T)^2] = E[(X - T)^2 | X < T]\Pr[X < T] \quad (4.21\,b)$$

(It is assumed that $\Pr[X = T] = 0$). Note that the factors $\Pr[X > T]$, $\Pr[X < T)$ make allowance for how often positive and negative deviations from T occur.

The multiplier $1/(3\sqrt{2})$, while the earlier PCIs use $\frac{1}{3}$, arises from the fact that for a *symmetrical* distribution with variance

σ^2 and expected value T we would have

$$E_{X>T}[(X-T)^2] = E_{X<T}[(X-T)^2] = \tfrac{1}{2}\sigma^2 \qquad (4.22)$$

Finally they define

$$C_{\mathrm{jkp}} = \min\left(CU_{\mathrm{jkp}}, CL_{\mathrm{jkp}}\right) = \frac{1}{3\sqrt{2}}\min\left(\frac{\mathrm{USL}-T}{\{E_{X>T}[(X-T)^2]\}^{\frac{1}{2}}},\right.$$

$$\left.\frac{T-\mathrm{LSL}}{\{E_{X<T}[(X-T)^2]\}^{\frac{1}{2}}}\right) \qquad (4.23)$$

Note that if we have $T=\tfrac{1}{2}(\mathrm{USL}+\mathrm{LSL})=m$, so that $\mathrm{USL}-T=T-\mathrm{LSL}=d$, then

$$C_{\mathrm{jkp}} = \frac{d}{3\sqrt{2}}\max\left([E_{X>T}[(X-T)^2], E_{X<T}[(X-T)^2]]\right)^{-\frac{1}{2}}$$

$$(4.24)$$

4.4.1 Estimation of C_{jkp}

A natural estimator of C_{jkp} is

$$\hat{C}_{\mathrm{jkp}} = \frac{1}{3\sqrt{2}}\min\left(\frac{\mathrm{USL}-T}{(S_+/n)^{\frac{1}{2}}}, \frac{T-\mathrm{LSL}}{(S_-/n)^{\frac{1}{2}}}\right) \qquad (4.25\,a)$$

where

$$S_+ = \sum_{X_i>T}(X_i-T)^2 \quad \text{and} \quad S_- = \sum_{X_i<T}(X_i-T)^2$$

Note that

$$E[S_+] = nE[(X-T)^2|X>T]\mathrm{Pr}[X>T]$$

so that S_+/n (*not* $S_+/[$number of X_is greater than $T]$) is an unbiased estimator of (4.21 *a*) (and S_-/n) is an unbiased estimator of (4.21 *b*). The estimator $\hat{C}_{jkp}$ can be calculated quite straightforwardly – the only special procedure needed is separate calculation of sums of squares for $X_i > T$ and $X_i < T$, as for Boyles' modified C_{pm} index, described in section 3.5.

Our analysis will be based on the very reasonable condition LSL $< T <$ USL. We can express $\hat{C}_{jkp}$ as

$$\hat{C}_{jkp} = \frac{d\sqrt{n}}{(3\sqrt{2})\sigma} \min\left(\frac{\text{USL} - T}{d\left(\frac{S_+}{\sigma^2}\right)^{\frac{1}{2}}}, \frac{T - \text{LSL}}{d\left(\frac{S_-}{\sigma^2}\right)^{\frac{1}{2}}}\right) \quad (4.25\,b)$$

where σ is an arbitrary constant.

To study the distribution of $\hat{C}_{jkp}$ it will be convenient to consider the statistic

$$\hat{D} = \frac{n}{18}\left(\frac{d}{\sigma}\right)^2 \hat{C}_{jkp}^{-2} = \max(a_1 S_+ \sigma^{-2}, a_2 S_- \sigma^{-2}) \quad (4.26)$$

where

$$a_1 = \left(\frac{\text{USL} - T}{d}\right)^{-2} \quad \text{and} \quad a_2 = \left(\frac{T - \text{LSL}}{d}\right)^{-2}$$

Note that $a_1^{-\frac{1}{2}} + a_2^{-\frac{1}{2}} = 2$, and

$$\text{E}[\hat{C}_{jkp}^r] = \left\{\frac{n}{18}\left(\frac{d}{\sigma}\right)^2\right\}^{\frac{1}{2}r} \text{E}[\hat{D}^{-\frac{1}{2}r}]$$

The distribution of $\hat{D}$ will be, in general, quite complicated. In Appendix 4A we discuss a special case in which the distribution of X is, indeed, normal with expected value T and variance σ^2. Although this is not, in fact, an asymmetrical distribution, consideration of this case can provide an initial point of reference. Later we will indicate ways in which the results can be extended to somewhat broader situations – though not as broad as we would wish.

Table 4.10 presents numerical values of

$$E[\hat{C}_{jkp}/C_{jkp}] \text{ and S.D. } (\hat{C}_{jkp}/C_{jkp})$$

(assuming normal process distribution) for several values of $(USL - T)/d$, and $n = 10(5)50$. Note that if $(USL - T)/d = 1$, then $T = (LSL + USL)/2$.

It is instructive to compare these values with similar quantities for $\hat{C}_{pk}/C_{pk}$ and $\hat{C}_{pm}/C_{pm}$, in Kotz and Johnson (1992) and Pearn *et al.* (1992) respectively (see also Tables 2.4 and 3.1).

In general the estimator $\hat{C}_{jkp}$ is biased. The bias is negative when $T = \frac{1}{2}(USL + LSL)$ but increases as $(USL - T)/d$ decreases. It is quite substantial when $(USL - T)/d$ is as small as 0.4. As might be expected the variance of $\hat{C}_{jkp}$ decreases as n increases – it increases as $(USL - T)/d$ decreases, as the target value gets nearer to the upper specification limit. The bias, also, decreases as n increases; this effect is particularly noticeable for smaller values of $(USL - T)/d$. For $n \geqslant 25$, and $(USL - T)/d \leqslant 0.60$ the stability of both E and S.D. is noteworthy.

Since the index C_{jkp} is intended to make allowance for asymmetry in the distribution of the process characteristic X it is of interest to learn something of the distribution of the estimator $\hat{C}_{jkp}$ under such conditions. The analysis in Appendix 4A can be extended to certain kinds of asymmetry of distribution of X.

We note two ways in which this may be done.

1. If the population density function of X is

$$\tfrac{1}{2}[g(x; T, \sigma_1) + g(-x: -T, \sigma_2)] \qquad (4.27)$$

Table 4.10 Values of $E = E[\hat{C}_{jkp}]/C_{jkp}$ and S.D. = S.D. $(\hat{C}_{jkp})/C_{jkp}$, when $\xi = T$.

$(USL-T)/d =$		0.10	0.20	0.30	0.40	0.50	0.60	0.70	0.80	0.90	1.00
$n=$											
10	E.	1.345	1.314	1.289	1.263	1.232	1.193	1.143	1.082	1.008	0.921
	S.D.	1.159	0.870	0.732	0.631	0.545	0.466	0.395	0.332	0.283	0.250
15	E.	1.175	1.172	1.169	1.163	1.153	1.135	1.107	1.063	1.000	0.918
	S.D.	0.556	0.511	0.480	0.449	0.412	0.369	0.320	0.270	0.227	0.198
20	E.	1.117	1.117	1.117	1.115	1.112	1.104	1.086	1.054	0.999	0.919
	S.D.	0.374	0.368	0.361	0.352	0.336	0.312	0.278	0.237	0.197	0.170
25	E.	1.089	1.089	1.089	1.088	1.087	1.083	1.072	1.048	0.999	0.922
	S.D.	0.298	0.297	0.296	0.293	0.287	0.273	0.250	0.215	0.178	0.152
30	E.	1.072	1.072	1.072	1.072	1.071	1.069	1.062	1.043	1.000	0.924
	S.D.	0.256	0.256	0.255	0.255	0.252	0.244	0.228	0.199	0.164	0.139
35	E.	1.060	1.060	1.060	1.060	1.060	1.059	1.054	1.039	1.001	0.927
	S.D.	0.228	0.228	0.228	0.227	0.226	0.222	0.211	0.187	0.153	0.129
40	E.	1.052	1.052	1.052	1.052	1.052	1.051	1.048	1.036	1.002	0.930
	S.D.	0.207	0.207	0.207	0.207	0.207	0.204	0.196	0.176	0.145	0.121
45	E.	1.045	1.045	1.045	1.045	1.045	1.045	1.043	1.034	1.003	0.932
	S.D.	0.192	0.192	0.192	0.192	0.191	0.190	0.184	0.168	0.138	0.115
50	E.	1.041	1.041	1.041	1.041	1.041	1.040	1.039	1.032	1.003	0.934
	S.D.	0.179	0.179	0.179	0.179	1.079	0.178	0.174	0.160	0.132	0.109

where $g(x; T; \sigma)$ is the PDF of the half-normal distribution:

$$g(x; T, \sigma) = \begin{cases} \dfrac{2}{\sigma\sqrt{2\pi}} \exp\left\{ -\dfrac{1}{2}\left(\dfrac{x-T}{\sigma}\right)^2 \right\} & \text{for } x \geq T \\ 0 & \text{for } x < T \end{cases}$$

then the distribution of $\hat{D}$ is as in (4.26) with a_j replaced by $a_j(\sigma/\sigma_j)^2 (j = 1, 2)$.

Here σ^2 is the variance of the process distribution, namely

$$\sigma^2 = \frac{1}{2}(\sigma_1^2 + \sigma_2^2) - \frac{1}{2\pi}(\sigma_1 - \sigma_2)^2$$

The use of distributions of form (4.27) in quality control has been suggested by Barnard (1989) who ascribes them to Fechner (1897). Rudolff and Hoffmann (1990) found that a distribution of this form would fit an observed distribution of textile strength quite well.

2. The number, K of X_is exceeding T can have a binomial distribution with parameters n, p with $0 < p < 1$.

If (1.) and (2.) are combined, the density function of X will be

$$p g(x; T, \sigma_1) + (1 - p)g(-x; -T, \sigma_2) \tag{4.28}$$

For either (1.) or (2.) the expected value of X is not, in general, T for density functions (4.27) or (4.28). (The distribution of X is a mixture of a half-normal and a negative half-normal). However, the conditional distributions of

$$\frac{\displaystyle\sum_{X_i > T} (X_i - T)^2}{\sigma_1^2} \quad \text{and} \quad \frac{\displaystyle\sum_{X_i < T} (X_i - T)^2}{\sigma_2^2},$$

given K, are still χ_K^2 and χ_{n-K}^2, respectively, as noted in Appendix 4A.

4.5 ASYMPTOTIC PROPERTIES

There are some large-sample properties of PCIs which apply to a wide range of process distributions and so contribute to our knowledge of behaviour of PCIs under non-normal conditions. Utilization of some of these properties calls for knowledge of the shape factors λ_3 and λ_4, and the need to estimate the values of these parameters can introduce substantial errors, as we have already noted. Nevertheless, asymptotic properties, used with clear understanding of their limitations, can provide valuable insight into the nature of indices.

Large-scale approximations to the distributions of $\hat{C}_p$, $\hat{C}_{pk}$ and $\hat{C}_{pm}$ have been studied by Chan *et al.* (1990).

Since

$$\hat{C}_p^{-1} = \left(\frac{3}{d}\right)\hat{\sigma}$$

the asymptotic distribution of $\hat{C}_p$ can easily be approached by considering the asymptotic distribution of $\hat{\sigma}$.

Chan *et al.* (1990) provided a rigorous proof that $\sqrt{n}(\hat{C}_p - C_p)$ has an asymptotic normal distribution with mean zero and variance

$$\sigma_p^2 = \tfrac{1}{4}(\beta_2 - 1)C_p^2 \tag{4.29}$$

If $\beta_2 = 3$ (as for normal process distribution), $\sigma_p^2 = \tfrac{1}{2}C_p^2$.

For $\hat{C}_{pk}$, Chan *et al.* (1990) found that if $\xi \neq m$ the asymptotic distribution of $\sqrt{n}(\hat{C}_{pk} - C_{pk})$ is normal with expected value zero and variance

$$\sigma_{pk}^2 = \frac{1}{9} + \frac{\varphi}{2}\sqrt{\beta_1}\, C_{pk} + \frac{1}{4}(\beta_2 - 1)C_{pk}^2 \tag{4.30}$$

where $\varphi = 1$ if $\xi > m$, $\varphi = -1$ if $\xi < m$.

If the process distribution is normal, then $\sqrt{\beta_1} = 0$ and

$\beta_2 = 3$, so that

$$\sigma_{pk}^2 = \tfrac{1}{9} + \tfrac{1}{2} C_{pk}^2 \qquad (4.31)$$

In the very special case when $\xi = m$, the limiting distribution of $\sqrt{n}(\tilde{C}_{pk} - C_{pk})$ is that of

$$-\tfrac{1}{3}\{|U_1| + \tfrac{1}{2} C_{pk}(\beta_2 - 1)^{\frac{1}{2}} U_2\} \qquad (4.32)$$

(note that $C_p = C_{pk}$ in this case) where U_1 and U_2 are standardized bivariate normal variables, with correlation coefficient $\sqrt{\beta_1}/(\beta_2 - 1)^{\frac{1}{2}}$. If the process distribution is normal the distribution is that of

$$-\frac{1}{3}\left\{|U_1| + \frac{1}{\sqrt{2}} C_{pk} U_2\right\} \qquad (4.33)$$

where U_1 and U_2 are independent standardized normal variables.

For $\hat{C}_{pm}$, we have $\sqrt{n}(\hat{C}_{pm} - C_{pm})$ asymptotically normally distributed with expected value zero and variance

$$\sigma_{pm}^2 = \frac{\left(\dfrac{\xi - m}{\sigma}\right)^2 + \sqrt{\beta_1}\left(\dfrac{\xi - m}{\sigma}\right) + \dfrac{1}{4}(\beta_2 - 1)}{\left\{1 + \left(\dfrac{\xi - m}{\sigma}\right)^2\right\}^2} C_{pm}^2 \qquad (4.34)$$

If the process distribution is normal, $\sqrt{\beta_1} = 0$ and $\beta_2 = 3$, hence

$$\sigma_{pm}^2 = \frac{\left(\dfrac{\xi - m}{\sigma}\right)^2 + \dfrac{1}{2}}{\left\{1 + \left(\dfrac{\xi - m}{\sigma}\right)^2\right\}^2} C_{pm}^2 \qquad (4.35)$$

and if, also, $\xi = m$

$$\sigma^2_{pm} = \tfrac{1}{2} C^2_{pm} \tag{4.36}$$

(In this case, also, $C_p = C_{pm}$).

APPENDIX 4.A

As noted in the text, it is assumed that the distribution of X is normal, with expected value T and standard deviation σ.

With the stated assumptions, we know that:

1. the distribution of $(X_i - T)^2$ is that of $\chi^2_1 \sigma^2$;
2. this is also the conditional distribution of $(X_i - T)^2$, whether $X_i > T$ or $X_i < T$;
3. the number, K, of Xs which exceed T has a binomial distribution with parameters $n, \tfrac{1}{2}$ – denoted $\text{Bin}(n, \tfrac{1}{2})$; and hence
4. given K, the conditional distributions of $S_+ \sigma^{-2}$ and $S_- \sigma^{-2}$ are those of χ^2_K, χ^2_{n-K} respectively, and $S_+ \sigma^{-2}$ and $S_- \sigma^{-2}$ are mutually independent; and also
5. the distribution of $H = S_+/(S_+ + S_-)$ is that of $\chi^2_K/(\chi^2_K + \chi^2_{n-K})$ which is $\text{Beta}(\tfrac{1}{2}K, \tfrac{1}{2}(n-K))$ so that the density function of H is

$$f_H(h) = \{B(\tfrac{1}{2}K, \tfrac{1}{2}(n-K))\}^{-1} h^{\frac{1}{2}K-1} (1-h)^{\frac{1}{2}(n-K)-1} \quad 0 < h < 1 \tag{4.37}$$

for $K = 1, 2, \ldots, n-1$, and H and $S_+ + S_-$ are mutually independent; and further
6. the conditional distributions of $S_+ \sigma^{-2}$ and $S_- \sigma^{-2}$, given K and H, are those of $H\chi^2_n$ and $(1-H)\chi^2_n$, respectively.

From (4.26)

$$
\hat{D} = \begin{cases} a_1 S_+ \sigma^{-2} & \text{for } \dfrac{S_+}{S_1} > \dfrac{a_2}{a_1} \\[3mm] a_2 S_- \sigma^{-2} & \text{for } \dfrac{S_+}{S_-} < \dfrac{a_2}{a_1} \end{cases} \tag{4.38}
$$

So $\hat{D}$ is distributed as

$$
\begin{cases} a_1 H \chi_n^2 & \text{for } \dfrac{H}{(1-H)} > \dfrac{a_2}{a_1} \text{ i.e. } H > \dfrac{a_2}{a_1 + a_2} \\[3mm] a_2 (1-H) \chi_n^2 & \text{for } H < \dfrac{a_2}{a_1 + a_2} \end{cases} .
$$

The overall distribution of $\hat{D}$ can be represented as

$$
\hat{D} \sim \begin{cases} a_1 H & \text{for } H > b \\ a_2 (1-H) & \text{for } H < b \end{cases} \chi_n^2
$$

$$
\bigwedge_H \text{Beta}(\tfrac{1}{2}K, \tfrac{1}{2}(n-K)) \bigwedge_K \text{Bin}(n, \tfrac{1}{2}) \tag{4.39}
$$

where the symbol $\bigwedge_Y$ means 'mixed with respect to Y' having the distribution that follows, and $b = a_2/(a_1 + a_2)$.

Without loss of generality, we assume that $USL - T \leqslant T - LSL$. Then, for a symmetrical distribution with mean T and variance σ^2,

$$
C_{jkp} = \frac{d}{3\sigma} \frac{USL - T}{d} = \frac{d}{3\sigma} \frac{1}{\sqrt{a_1}} \tag{4.40}
$$

and

$$
\frac{1}{\sqrt{a_1}} + \frac{1}{\sqrt{a_2}} = 2
$$

Hence (from (4.26) and 4.40))

$$\frac{\hat{C}_{jkp}}{C_{jkp}} = \left(\frac{na_1}{2}\right)^{\frac{1}{2}} \hat{D}^{-\frac{1}{2}}$$

(4.41)

and

$$E\left[\left(\frac{\hat{C}_{jkp}}{C_{jkp}}\right)^r\right] = \left(\frac{na_1}{2}\right)^{\frac{1}{2}r} E[\hat{D}^{-\frac{1}{2}r}]$$

(4.42)

Now, from (4.37), (4.38) and (4.39), noting that

$$E[(\chi_n^2)^{-\frac{1}{2}r}] = \{2^{\frac{1}{2}r}\Gamma(\tfrac{1}{2}n)\}^{-1}\Gamma(\tfrac{1}{2}(n-r))$$

$$E[\hat{D}^{-\frac{1}{2}r}] = \frac{\Gamma(\tfrac{1}{2}(n-r))}{2^{\frac{1}{2}r}\Gamma(\tfrac{1}{2}n)} \frac{1}{2^n}\left[a_2^{-\frac{1}{2}r} + \sum_{k=1}^{n-1} \frac{\binom{n}{k}}{B(\tfrac{1}{2}k, \tfrac{1}{2}(n-k))}\right.$$

$$\times \{a_1^{-\frac{1}{2}r}B_{1-b}(\tfrac{1}{2}(n-k), \tfrac{1}{2}(k-r))$$

$$\left. + a_2^{-\frac{1}{2}r}B_b(\tfrac{1}{2}k, \tfrac{1}{2}(n-k-r))\} + a_1^{-\frac{1}{2}r}\right]$$

whence

$$E\left[\left(\frac{\hat{C}_{jkp}}{C_{jkp}}\right)^r\right] = \frac{n^{\frac{1}{2}r}\Gamma(\tfrac{1}{2}(n-r))}{2^{n+r}\Gamma(\tfrac{1}{2}n)}\left[\left(\frac{a_1}{a_2}\right)^{\frac{1}{2}r} + 1\right.$$

$$+ \sum_{k=1}^{n-1} \frac{\binom{n}{k}}{B(\tfrac{1}{2}k, \tfrac{1}{2}(n-k))}\left\{B_{1-b}(\tfrac{1}{2}(n-k), \tfrac{1}{2}(k-r))\right.$$

$$\left.\left. + \left(\frac{a_1}{a_2}\right)^{\frac{1}{2}r} B_b(\tfrac{1}{2}k, \tfrac{1}{2}(n-k-r))\right\}\right]$$

(4.43)

where

$$B_v(u_1, u_2) = \int_0^v y^{u_1 - 1}(1 - y)^{u_2 - 1} \, dy$$

and $B(u_1, u_2) = B_1(u_1, u_2)$ (see section 1.6).
In particular

$$E\left[\frac{\hat{C}_{jkp}}{C_{jkp}}\right] = \frac{\Gamma(\frac{1}{2}(n-1))\sqrt{n}}{2^{n+1}\Gamma(\frac{1}{2}n)}\left[\left(\frac{a_1}{a_2}\right)^{\frac{1}{2}} + 1\right.$$

$$+ \sum_{k=1}^{n-1} \frac{\binom{n}{k}}{B(\frac{1}{2}k, \frac{1}{2}(n-k))}\left\{B_{1-b}(\frac{1}{2}(n-k), \frac{1}{2}(k-1))\right.$$

$$\left.+ \left(\frac{a_1}{a_2}\right)^{\frac{1}{2}r} B_b(\frac{1}{2}k, \frac{1}{2}(n-k-r))\right\}\right]$$

BIBLIOGRAPHY

Balitskaya, E.O. and Zolotuhina, L.A. (1988) On the representation of a density by an Edgeworth series, *Biometrika*, **75**, 185–187.

Barnard, G.A. (1989) Sophisticated theory and practice in quality improvement, *Phil. Trans. R. Soc.*, London **A327**, 581–9.

Boyles, R.A. (1991). The Taguchi capability index, *J. Qual. Technol.*, 23, 17–26.

Chan, L.K., Cheng, S.W. and Spiring, F.A. (1988). The robustness of process capability index C_p to departures from normality. In *Statistical Theory and Data Analysis*, II (K. Matusita, ed.), North-Holland, Amsterdam, 223–9.

Chan, L.K., Xiong, Z. and Zhang, D. (1990) On the asymptotic distributions of some process capability indices, *Commun. Statist. – Theor. Meth.*, **19**, 1–18.

Clements, J.A. (1989). Process capability calculations for non-normal calculations, *Qual. Progress*, **22**(2), 49–55.

Efron, B. (1982) *The Jackknife, the Bootstrap and Other Re-sampling Plans*, SIAM, CBMS-NSF Monograph, **38**, SIAM: Philadelphia, Pennsylvania.

English, J.R. and Taylor, G.D. (1990) Process Capability Analysis – A Robustness Study, MS, Dept. Industr. Eng., University of Arkansas, Fayetteville.

Fechner, G.T. (1897) *Kollektivmasslehre*, Engelmann, Leipzig.

Franklin, L.A. and Wasserman, G. (1992). Bootstrap confidence interval estimates of C_{pk}: An Introduction, *Commun. Statist.-Simul. Comp.*, **20**, 231–42.

Gruska, G.F., Mirkhani, K. and Lamberson, L.R. (1989) *Non-normal Data Analysis*, Applied Computer Solutions, Inc.; St. Clare Shores, Michigan.

Guenther, W.H. (1985). Two-sided distribution-free tolerance intervals and accompanying sample size problems, *J. Qual. Technol.*, **17**, 40–3.

Gunst, R.F. and Webster, J.T. (1978). Density functions of the bivariate chi-squared distribution, *J. Statist. Comp. Simul.*, **2**, 275–88.

Gunter, B.H. (1989) The use and abuse of C_{pk}, 2/3, *Qual. Progress*, **22**(3), 108–109; (5), 79–80.

Hall, P. (1992) Private communication,

Hall, P. and Martin, M.A. (1988) On bootstrap resampling and iteration, *Biometrika*, **75**, 661–71.

Hall, P., Martin, M.A. and Schucany, W.R. (1989) Better non-parametric bootstrap confidence intervals for the correlation coefficient, *J. Statist. Comp. Simul*, **33**, 161–72.

Hsiang, T.C. and Taguchi, G. (1985) A Tutorial on Quality Control and Assurance – The Taguchi Methods, *ASA Annual Meeting*, Las Vegas, Nevada.

Hung, K. and Hagen D. (1992) Statistical Computation Using GAUSS: Examples in Process Capability Research, *Tech. Rep.* Western Washington University, Bellingham.

Johnson, M. (1992) Statistics simplified, *Qual. Progress*, **25**(1), 10–11.

Johnson, N.L., and Korz, S. (1970) *Distributions in Statistics: Continuous Univariate Distributions*, John Wiley, New York.

Johnson, N.L., Kotz, S. and Pearn, W.L. (1992) Flexible process capability indices (Submitted for publication).

Kane, V.E. (1986) Process capability indices, *J. Qual. Technol.*, **18**, 41–52.

Kocherlakota, S., Kocherlakota, K. and Kirmani, S.N.U.A. (1992) Process capability indices under non-normality, *Internat. J. Math. Statist.* **1**.

Kotz, S. and Johnson, N.L. (1993) Process capability indices for non-normal populations, *Internat. J. Math. Statist.* **2**.

Marcucci, M.O. and Beazley, C.C. (1988) Capability indices: Process performance measures, *Trans. ASQC Tech. Conf., Dallas, Texas*, 516–22.

McCoy, P.F. (1991) Using performance indexes to monitor production processes, *Qual. Progress*, **24**(2), 49–55.

Munechika, M. (1986) Evaluation of process capability for skew distributions, *30th EOQC Conf., Stockholm, Sweden*.

Munechika, M. (1992) Studies on process capability in matching processes, *Mem. School Sci. Eng., Waseda Univ. (Japan)*, **56**, 109–124.

Pearn, W.L. and Kotz, S. (1992) Application of Clements' method for calculation second and third generation PCIs from non-normal Pearsonian populations. (Submitted for publication).

Pearn, W.L., Kotz, S. and Johnson, N.L. (1992) Distributional and inferential properties of process capability indices, *J. Qual. Technol.* **24**, 216–31.

Pearson, E.S. and Tukey, J.W. (1965) Approximate means and standard deviations based on differences between percentage points of frequency curves, *Biometrika*, **52**, 533–46.

Price, B. and Price, K. (1992) Sampling variability in capability indices, *Tech. Rep.* Wayne State University, Detroit, Michigan.

Rudolff, E. and Hoffman, L. (1990) Bicomponare Verteilung – eine erweiterte asymmetrische form der Gaußschen Normalverteilung, *Textiltechnik*, **40**, 49–500.

Schenker, N. (1985) Qualms about bootstrap confidence intervals, *J. Amer. Statist. Assoc.*, **80**, 360–1.

Subrahmaniam, K. (1966, 1968a, 1968b). Some contributions to the theory of non-normality, I; II; III. *Sankhya*, **28A**, 389–406; **30A**, 411–32; **30B**, 383–408.

5

Multivariate process capability indices

5.1 INTRODUCTION

Frequently – indeed usually – manufactured items need values of several different characteristics for adequate description of their quality. Each of a number of these characteristics must satisfy certain specifications. The assessed quality of the product depends, *inter alia*, on the combined effect of these characteristics, rather than on their individual values. With modern monitoring devices, simultaneous recording of several characteristics is becoming more feasible, and the utilization of such measurements is an important issue. Multivariate control methods (see Alt (1985) for a comprehensive review), although originating with the work of Hotelling (1947), have only recently become an active and fruitful field or research.

The use of PCIs in connection with multivariate measurements is hedged around with even more cautions and drawbacks than is the case for univariate measurements. In particular, the intrinsic difficulties arising from use of a single index as a quality measure are increased when the single index has to summarize measurements on several characteristics rather than just one. In fact, most of the multivariate PCIs which have been proposed, and will be discussed in this chapter, can be best understood as being univariate PCIs,

179

based on a particular function of the variables representing the characteristics. Further, so far as we are aware, multivariate PCIs are, as yet, used very rarely (if at all) in many industries.

The contents of this chapter should, therefore, be regarded mainly as theoretical background, indicating some interesting possibilities, and only an initial guide to practice. Nevertheless, we believe that study of this chapter will not prove to be a barren exercise. The reader should gain insight into the nature and consequences of multiple variation, and also to understand pitfalls in the simultaneous use of separate PCIs, one for each measured variate.

5.2 CONSTRUCTION OF MULTIVARIATE PCIs

Just as in the univariate case, we need to consider the inherent structure of variation of the measured characteristics – but, this time, not only variances but also correlations need to be taken into account. Deviations from target vector (**T**) also need to be considered.

The structure of variation has to be related to the specification region (R) for the measured variates $X_1, \ldots, X_v$. In principle this region might be of any form – later we will discuss a favourite theoretical form – though in practice it is usually in the form of a rectangular paralellopiped (a v-dimensional 'box') defined by

$$\mathrm{LSL}_i \leqslant X_i \leqslant \mathrm{USL}_i \quad i = 1, \ldots, v \qquad (5.1)$$

If the specification region is of this form, Chan *et al.* (1990) suggest using the product of the v univariate C_{pm} values as a multivariate PCI. A moments' reflection however, shows that apart from the serious defects of the univariate C_{pm}, described in Chapter 3, this could lead to absurd situations, even if the v measured variates are mutually independent. The reason for this is that a very bad (i.e. small) C_{pm} value for one variate can be compensated by sufficiently large C_{pm} values for the other values, giving an unduly favourable value for the multivariate PCI. (If one component is only rarely within specification limits, it is small consolation if all the others are never nonconforming!) A more promising line of attack may be possible using results of Kocherlakota and Kocherlakota (1991), who have derived the joint distribution of $C_{pm}s$ calculated for two characteristics, with a joint bivariate normal distribution. We will return to this work in section 5.5.

As with univariate PCIs, there are two possibe main lines of approach – one based on expected proportion of NC items, the other based primarily on loss functions. Both are based on assumptions: about distributional form in the first approach; and about form of loss function in the second. These more or less arbitrary assumptions are, of necessity, more extensive in the multivariate case, just because there are more variates involved. Note that the approaches of Lam and Littig (1992) and Wierda (1992a) (see chapter 2, section 2) can be extended straightforwardly to multivariate situations.

Corresponding to the assumption of normality in the univariate case, we have, in the multivariate case, the assumption of multinormality, with

$$F[X_i] = \xi_i$$
$$\text{var}(X_i) = \sigma_i^2$$
$$\text{corr}(X_i, X_{i'}) = \rho'_{ii} \quad i, i' = 1, \dots, v$$

i.e. expected value vector $\boldsymbol{\xi} = (\xi_1, \xi_2, \dots, \xi_v)$ and variance–covariance matrix $\mathbf{V}_0 = (\rho_{ii'}\sigma_i\sigma'_i)$ with $\rho_{ii} = 1$ for all i. This means that the joint PDF of $\mathbf{X} = (X_1, \dots, X_v)$ is

$$f_{\mathbf{X}}(\mathbf{x}) = (2\pi)^{-v/2}|\mathbf{V}_0|^{-\frac{1}{2}}\exp\{-\tfrac{1}{2}(\mathbf{x} - \boldsymbol{\xi})'\mathbf{V}_0^{-1}(\mathbf{x} - \boldsymbol{\xi})\} \quad (5.2)$$

(see section 1.8.)

If a loss-function approach is employed, a natural generalization of the univariate loss function $k(X-T)^2$ is the quadratic form

$$L(\mathbf{X}) = (\mathbf{X} - \mathbf{T})'\mathbf{A}^{-1}(\mathbf{X} - \mathbf{T}) \tag{5.3}$$

with a specification region $L(\mathbf{X}) \leqslant c^2$. ($\mathbf{A}^{-1}$ is called the generating matrix of this quadratic form). T. Johnson (1992) interprets c^2 as a maximum 'worth', attained when $\mathbf{X} = \mathbf{T}$ – i.e. all characteristics attain their respective target values – and $L(\mathbf{X})$ is a loss of 'worth', so that zero worth is reached at the boundary of R.

Note that $\mathbf{A}$ does not have any necessary relationship with the variance covariance matrix $\mathbf{V}_0$. It is, indeed, very unlikely that $\mathbf{A}$ would be identical with $\mathbf{V}_0$, (though tempting for theoreticians to explore what would be the consequences if this were the case, as we will see later).

In the following sections we describe a few proposed multivariate PCIs and provide some critical assessment of their properties.

Before concluding this section however, we stress that any index of process capability, based on multivariate characteristics, that is a single number, has an even higher risk of being misused and misinterpreted than is the case for univariate PCIs. The time-honoured measures and techniques of classical statistical multivariate analysis – e.g. Hotelling's T^2, Mahalanobis' distance, principal component analysis – provide ways of obtaining more detailed assessment of process variation. It would be inefficient, not to utilize appropriate forms of these well-established measures and techniques. Reduction to a single statistic, however ingeniously constructed, is equivalent to replacing a multivariate problem by a univariate one, with attendant loss of information.

However, if a process is stable, it is much easier to monitor a single index than, for example to use many mean and range control charts. Nevertheless, a vector-valued multivariate PCI has been proposed. This will be discussed, briefly, in section 5.6.

5.3 MULTIVARIATE ANALOGS OF C_p

If we accept the assumption of multinormality (5.2), it would be natural to use, as a basis for construction of PCIs, the quadratic form

$$W = (\mathbf{X} - \xi)\mathbf{V}_0^{-1}(\mathbf{X} - \xi) \tag{5.4}$$

The statistic W would have a (central) chi-squared distribution with v degrees of freedom. If, also (see remarks near the end of section 5.2), the specification region R were of form

$$W \leqslant \chi_{v,0.9973}^2 = c_v^2$$

the expected proportion of NC product would be 0.27%.

We also note that it has been suggested that an estimate ($\hat{p}$, say) of the expected proportion of NC items based on observed data, and exploiting the assumed form of process distribution (usually multinormality), might be, itself, used as a PCI. We have already noted this suggestion for the univariate case, in section 2.1 It is important to realize that the dependence on correct form of process distribution is even heavier in the multivariate than in the univariate case. Littig *et al.* (1992) suggest using $\Phi^{-1}(\frac{1}{2}(\hat{p}+1))$ as a PCI (as in the univariate case). See also end of section 5.5.

Recall that the univariate index C_p was defined as

$$C_p = \frac{\text{length of specification interval}}{6 \times (\text{standard deviation of } X)} \tag{5.5}$$

the factor 6 being used because, if $C_p = 1$ and variation is normal, it is just possible – by making $\xi = \frac{1}{2}(\text{USL} + \text{LSL})$ – to have the expected proportion of NC product as small as 0.27%. Taam *et al.* (1991), regarding the denominator in (5.5) as 'length of central interval containing 99.73% of values of X', propose the natural generalization

$$C_p = \frac{\text{volume of specification region}}{\text{volume of region containing 99.73\% of values of } X.}$$

$$(5.6)$$

The denominator is the volume of the ellipsoid

$$(\mathbf{x} - \xi)' \mathbf{V}_0^{-1} (\mathbf{x} - \xi) \leqslant \chi^2_{v, 0.9973} \tag{5.7}$$

which is

$$\frac{(\pi \chi^2_{v, 0.9973})^{\frac{1}{2}v}}{\Gamma(\frac{1}{2}v + 1)} |\mathbf{V}_0|^{\frac{1}{2}} \tag{5.8}$$

[Of course, this would be the volume of the ellipsoid (5.7), whatever the value of ξ. The ellipsoid would contain 99.73% of values of X however, *only* with ξ equal to the expected value vector.]

This leads us to the definition

$$C_p = \frac{\text{volume of } R}{a_v |\mathbf{V}_0|^{\frac{1}{2}}} \tag{5.9}$$

where $a_v = (\pi \chi^2_{v, 0.9973})^{\frac{1}{2}v} / \Gamma(\frac{1}{2}v + 1)$.

However, some modification of this definition is needed in order to provide a genuine generalization of the coverage property of $C_p(v = 1)$, which would be that, if $\xi = \mathbf{T}$ and $\mathbf{T}$ is the centre of the specification region, then $C_p = 1$ should ensure that 99.73% of values of $\mathbf{X}$ fall within R. This property is clearly not satisfied if the 'box' region $LSL_i \leqslant X_i \leqslant USL_i$ $(i = 1, \ldots, v)$ is used. The volume of the rectangular paralellopiped is

$$\prod_{i=1}^{v} (USL_i - LSL_i) \tag{5.10}$$

but $Pr[LSL_i \leqslant X_i \leqslant USL_i$ for all $i = 1, \ldots, v]$ is not necessarily equal to 99.73%, even if

$$\prod_{i=1}^{v} (USL_i - LSL_i) = \frac{(\pi \chi^2_{0.9973})^{\frac{1}{2}v} |\mathbf{V}_0|}{\Gamma(\frac{1}{2}v + 1)} \tag{5.11}$$

The property *would* hold if R were of form

$$(\mathbf{X} - \xi)'\mathbf{V}_0^{-1}(\mathbf{X} - \xi) \leqslant K^2 \tag{5.12a}$$

but, as we noted in section 5.2, this is unlikely to occur. Even if an ellipsoidal R, of form

$$(\mathbf{X} - \xi)'\mathbf{A}^{-1}(\mathbf{X} - \xi) \leqslant K^2 \tag{5.12b}$$

were specified, it is unlikely that $\mathbf{A}$ would equal $\mathbf{V}_0$.

Taam *et al.* (1991) propose to avoid this difficulty by using a 'modified specification region' – R^*, say – defined as the greatest volume ellipsoid with generating matrix $\mathbf{V}_0^{-1}$, contained within the actual specification region, R. If this ellipsoid is given by (5.12 a), then

$$\mathbf{C}_p^* = \frac{\text{volume of } R^*}{\text{volume of } (\mathbf{x} - \xi)'\mathbf{V}_0^{-1}(\mathbf{x} - \xi) \leqslant \chi^2_{v, 0.9973}} = \left(\frac{K^2}{\chi^2_{v, 0.9973}} \right)^{v/2} \tag{5.13}$$

for any ξ and any v.

The most one might reasonably hope for, in practice, is that the region R is defined by (5.12 b), and it is known that

$\mathbf{V}_0$ is proportional to $\mathbf{A}$, i.e. $\mathbf{V}_0 = \theta^2 \mathbf{A}$, though the value of the multiplier, θ^2, is not known. Total rejection of this possibility (an opinion held in some circles) seems to be an unduly pessimistic outlook, assuming that practitioners have so little knowledge of the process as to be unwilling to be somewhat flexible in adjusting engineering specifications to the actual behaviour of the process in order to benefit from available theory. We consider such an attitude to be unhealthy, and encourage practitioners to take advantage of knowledge of likely magnitude of correlations among characteristics in setting specification, if possible.

If $\mathbf{V}_0 = \theta^2 \mathbf{A}$ then

$$
\begin{aligned}
\mathbf{C}_p &= \frac{\text{volume of } \mathbf{x}'\mathbf{A}^{-1}\mathbf{x} \leqslant K^2}{\text{volume of } \mathbf{x}'\mathbf{V}_0^{-1}\mathbf{x} \leqslant \chi^2_{v,\,0.9973}} \\[2mm]
&= \frac{\text{volume of } \mathbf{x}'\mathbf{V}_0^{-1}\mathbf{x} \leqslant (K/\theta)^2}{\text{volume of } \mathbf{x}'\mathbf{V}_0^{-1}\mathbf{x} \leqslant \chi^2_{v,\,0.9973}} \\[2mm]
&= \left(\frac{K^2}{\theta^2 \chi^2_{v,\,0.9973}} \right)^{v/2}
\end{aligned}
\tag{5.14}
$$

Pearn *et al.* (1992) reach a similar PCI by regarding $\chi^2_{v,\,0.9973}$ as a generalized 'length' corresponding to a proportion 0.0027 of NC items, and K^2/h^2 as a generalized 'allowable length'. They define

$$
_v\mathbf{C}_p = \frac{K}{\theta \chi_{v,\,0.9973}} = \mathbf{C}_p^{1/v}
\tag{5.15}
$$

Estimation of $\mathbf{C}_p$ (or $_v\mathbf{C}_p$) is equivalent to estimation of θ. If $\mathbf{V}_0 = \theta^2 \mathbf{A}$, and values $\mathbf{X}_j = (x_{1j}, \ldots, x_{vj})$ are available for n individuals ($j = 1, \ldots, n$), the statistic,

$$
S = \sum_{j=1}^{n} (\mathbf{X}_j - \bar{\mathbf{X}})' \mathbf{A}^{-1} (\mathbf{X}_j - \bar{\mathbf{X}})
$$

is distributed as $\theta^2 \chi^2_{(n-1)v}$, and $S/\{(n-1)v\}$ is an unbiased estimator of θ^2.

A natural estimator of $_v C_p$ is

$$_v \hat{C}_p = \frac{K}{\chi_{v,0.9973}} \left(\frac{(n-1)v}{S} \right)^{\frac{1}{2}} \qquad (5.16\,a)$$

which is distributed as

$$\frac{K}{\chi_{v,0.9973}} \frac{((n-1)v)^{\frac{1}{2}}}{\theta \chi_{(n-1)v}} = \frac{((n-1)v)^{\frac{1}{2}}}{\chi_{(n-1)v}}\, {_v C_p}$$

and a $100(1-\alpha)\%$ confidence interval for $_v C_p$ is

$$\left(\frac{\chi_{(n-1)v,\alpha/2}}{\{(n-1)v\}^{\frac{1}{2}}}\, {_v \hat{C}_p}, \ \frac{\chi_{(n-1)v,1-\alpha/2}}{\{(n-1)v\}^{\frac{1}{2}}}\, {_v \hat{C}_p} \right) \qquad (5.16\,b)$$

(cf (2.9 c)).

5.4 MULTIVARIATE ANALOGS OF C_{pm}

Taam *et al.* (1991) proceed to define a mutivariate analog of C_{pm} by the formula

$$C^*_{pm} = \frac{\text{volume of } R^*}{\text{volume of ellipsoid } x'V_T^{-1}x < c^2} \qquad (5.17)$$

(cf. (5.13)) where

$$V_T = V_0 + (\xi - T)(\xi - T)'$$

Since

$$|V_T| = |V_0| \{ 1 + (\xi - T)' V_0^{-1} (\xi - T) \}^{\frac{1}{2}}$$

$$C^*_{pm} = C^*_p \{ 1 + (\xi - T)' V_0^{-1} (\xi - T) \}^{-\frac{1}{2}} \qquad \text{(cf. (3.8))(5.18)}$$

(Taam *et al.* (1991) use the notation MCP for C_{pm}^*). This index has the property that when the process mean vector ξ equals the target vector $\mathbf{T}$, and the index has the value 1, then 99.73% of the process values lie within the modified specification region R^*. This is analogous to the property of the univariate C_{pm} index.

The matrix $\mathbf{V_T}$ can be estimated unbiasedly as

$$\hat{\mathbf{V}}_{\mathbf{T}} = \frac{1}{n} \sum_{j=1}^{n} (\mathbf{X}_j - \mathbf{T})(\mathbf{X}_j - \mathbf{T})' \qquad (5.19\,a)$$

and $\mathbf{V_0}$ by

$$\hat{\mathbf{V}} = \frac{1}{n-1} \sum_{j=1}^{n} (\mathbf{X}_j - \bar{\mathbf{X}})(\mathbf{X}_j - \bar{\mathbf{X}})' \qquad (5.19\,b)$$

where

$$\bar{\mathbf{X}} = \frac{1}{n} \sum_{j=1}^{n} \mathbf{X}_j$$

Note that

$$\hat{\mathbf{V}}_{\mathbf{T}} = \frac{n-1}{n} \hat{\mathbf{V}} + (\bar{\mathbf{X}} - \mathbf{T})(\bar{\mathbf{X}} - \mathbf{T})' \qquad (5.19\,c)$$

Chan *et al.* (1990) assume that the specification region R is of form

$$(\mathbf{X} - \mathbf{T})'\mathbf{V}_0^{-1}(\mathbf{X} - \mathbf{T}) \leqslant K$$

i.e. the generating matrix of the specification region R is proportional to $\mathbf{V_0}$, the variance-covariance matrix of the measured variates. They define a multivariate analog of

C_{pm} as

$$C_{pm} = \left[\frac{v}{E[(X-T)'V_0^{-1}(X-T)]} \right]^{\frac{1}{2}}$$
$$= \{1 + v^{-1}(\xi-T)V_0^{-1}(\xi-T)\}^{-\frac{1}{2}} \qquad (5.20)$$

If V_0 is known, an unbiased estimator of the denominator of C_{pm}^2 is

$$\frac{1}{n} \sum_{j=1}^{n} (X_j-T)'V_0^{-1}(X-T)$$

and a natural estimator of C_{pm} is

$$\hat{C}_{pm} = \left[\frac{nv}{\sum_{j=1}^{n} (X_j-T)'V_0^{-1}(X_j-T)} \right]^{\frac{1}{2}} \qquad (5.21)$$

At this point we note that C_{pm} is subject to the same drawback as C_{pm}, noted in section 3.2. Identical values of C_{pm} can correspond to dramatically different expected proportions of NC items, if the specification region is not centered at T. Also, calculation of $\hat{C}_{pm}$ requires knowledge of the correct V_0. If an arbitrary matrix is used in place of V_0 in defining C_{pm}, we can encounter a similar effect of ambiguity in meaning of the value of C_{pm}. (See Chen (1992) and also Appendix 5.A). The numerator v is introduced because if $\xi = T$, then

$$E[(X-T)'V_0^{-1}(X-T)] = v$$

If $\xi = T$, then $\hat{C}_{pm}^{-2}$ is distributed as $(nv)^{-1}\chi_{nv}^2$. Hence, in these special circumstances

$$E[\hat{C}_{pm}] = \left(\frac{nv}{2} \right)^{\frac{1}{2}} \frac{\Gamma(\frac{1}{2}(nv-1))}{\Gamma(\frac{1}{2}nv)} \qquad (5.22\,a)$$

and

$$\text{var}(\hat{C}_{pm}) = \tfrac{1}{2}nv\left[\frac{\Gamma(\tfrac{1}{2}nv)\Gamma(\tfrac{1}{2}nv-1)-\{\Gamma(\tfrac{1}{2}(nv-1)\}^2}{\{\Gamma(\tfrac{1}{2}nv)\}^2}\right]$$

$$(5.22\,b)$$

(Chan *et al.* (1990)).

Percentage points for $\hat{C}_{pm}$ (on the assumption that $\xi = T$) are easily obtained since

$$\begin{aligned}
\Pr[\hat{C}_{pm} \leqslant G_\alpha] &= \Pr[\hat{C}_{pm}^{-2} \geqslant G_\alpha^{-2}]\\
&= \Pr[(nv)^{-1}\chi_{nv}^2 \geqslant G_\alpha^{-2}]\\
&= \Pr[\chi_{nv}^2 \geqslant nvG_\alpha^{-2}]
\end{aligned}$$

Hence, to make $\Pr[\hat{C}_{pm} \leqslant G_\alpha] = \alpha$ we take

$$G_\alpha = \frac{(nv)^{\frac{1}{2}}}{\chi_{nv,\,1-\alpha}} \tag{5.23}$$

Except when nv is small the approximation

$$G_\alpha \cong \frac{(2nv)^{\frac{1}{2}}}{z_{1-\alpha}+(2nv-1)^{\frac{1}{2}}} \tag{5.24\,a}$$

where $\Phi(z_{1-\alpha}) = 1-\alpha$, or even

$$G_\alpha \cong \left(\frac{z_{1-\alpha}}{(2nv)^{\frac{1}{2}}}+1-\frac{1}{4nv}\right)^{-1} \tag{5.24\,b}$$

give values which are sufficiently accurate for practical purposes. [For example, using (5.24 a)

when $nv = 40$ we find $G_{0.025} \cong 0.8245$ (correct value 0.8210) and $G_{0.05} \cong 0.8492$ (correct value 0.8470):

when $nv = 100$, we find $G_{0.025} \cong 0.8802$ (correct value 0.8785)

and $G_{0.05} \cong 0.8978$ (correct value 0.8968):

and when $nv = 200$, we find $G_{0.025} \cong 0.9118$ (correct value 0.9109)

and $G_{0.05} \cong 0.9251$ (correct value 0.9245)].

Pearn *et al.* (1992) felt that C_{pm} is not a true generalization of the univariate C_{pm}, and proposed the PCI

$$_v C_{pm} = {_v C_p} \left\{ 1 + \frac{1}{v} (\xi - T)' V_0^{-1} (\xi - T) \right\}^{-\frac{1}{2}}$$

$$= {_v C_p} \times C_{pm} \qquad (5.25)$$

(see (5.15) and (5.20)).

Note that the univariate indices C_{pm} and C_p are related by

$$C_{pm} = C_p \left\{ 1 + \left(\frac{\xi - T}{\sigma} \right)^2 \right\}^{-\frac{1}{2}}$$

where $\xi = E[X]$, $\sigma^2 = \text{var}(X)$ and T is the target value for X. A natural estimator of $_v C_{pm}$ is

$$_v \hat{C}_{pm} = \frac{K(nv)^{\frac{1}{2}}}{\theta \chi_{v, 0.9973}} \times \{ S + n(\bar{X} - T) V_0^{-1} (\bar{X} - T) \}^{-\frac{1}{2}} \quad (5.26)$$

In all the previous work, distribution theory has leaned heavily on the assumption that:

1. R is of form $(X - T)' A^{-1} (X - T) \leqslant K^2$ and
2. $A = V_0$, or, at least, A is proportional to V_0, i.e. $\theta^2 A = V_0$, as in (5.14).

In this second case, even with $\xi \neq T$, it is possible to evaluate the distribution of

$$\theta^2 \sum_{j=1}^{n} (X_j - T)' V_0^{-1} (X_j - T)$$

as that of

$$\theta^2 \times (\text{noncentral } \chi^2 \text{ with } nv \text{ degrees of freedom and}$$
$$\text{noncentrality parameter } n(\boldsymbol{\xi} - \mathbf{T})' \mathbf{V}_0^{-1} (\boldsymbol{\xi} - \mathbf{T}))$$

However, if $\mathbf{A}$ is not proportional to $\mathbf{V}_0$, the distribution of $(\mathbf{X} - \mathbf{T})' \mathbf{A}^{-1} (\mathbf{X} - \mathbf{T})$, although known, is complicated (Johnson and Kotz (1968)). (Appendix 5A outlines some details.)

Other approaches to measuring 'capability' from multivariate data include:

1. Use of the PCI

$$_v C_R = |\mathbf{V}_0 \mathbf{A}^{-1}| + (\boldsymbol{\xi} - \mathbf{T})' \mathbf{A}^{-1} (\boldsymbol{\xi} - \mathbf{T}) \qquad (5.27)$$

The motivation for this PCI is that it is expressed in terms of the matrix $\mathbf{A}$ used in the specification, rather than the process variance-covariance matrix – on the lines of Taam *et al.* (1991). If $v = 1$ and $\mathbf{A} = (\frac{1}{3}d)^2$ we have $_1 C_R = C_{pm}^{-2}$. Note that small values of $_v C_R$ are 'good', and large values are 'bad'.

2. Proceeding along the lines of Hsiang and Taguchi (1985) and Johnson (1992) (see section 3.2) we may introduce

$$L(\mathbf{x}) = (\mathbf{x} - \mathbf{T})' \mathbf{A}^{-1} (\mathbf{x} - \mathbf{T}) \qquad (5.28\,a)$$

as a loss function (generalizing $L(x) = k(x - T)^2$). We have

$$E[L(\mathbf{X})] = \text{tr}(\mathbf{A}^{-1} \mathbf{V}_0) + (\boldsymbol{\xi} - \mathbf{T})' \mathbf{A}^{-1} (\boldsymbol{\xi} - \mathbf{T}) \qquad (5.28\,b)$$

where $\text{tr}(\mathbf{M})$ denotes 'trace of $\mathbf{M}$', which is the sum of diagonal elements of $\mathbf{M}$. Large values of $E[L(\mathbf{X})]$ are, of course, 'bad', and small values, 'good'.

For readers' convenience we summarize, in Table 5.1, the scalar indices introduced in this chapter.

3. Chen (1992) has noted that a broad class of PCIs can be defined in the following way. Suppose the specification

region is

$$h(\mathbf{x} - \mathbf{T}) < r_0$$

Table 5.1 Scalar indices for multivariate data

Source	Symbol	Definition
Taam *et al.* (1991)	C_p	$\dfrac{\text{vol. specn. region}}{\text{vol. of } (\mathbf{x} - \xi)'\mathbf{V}_0^{-1}(\mathbf{x} - \xi) \leqslant \chi^2_{\nu, 0.9973}}$
Taam *et al.* (1991)	C_p^*	As C_p, with *modified* specn. region
Pearn *et al.* (1992)	$_\nu C_p$	$C_p^{1/\nu}$
Taam *et al.* (1991)	C_{pm}^*	As C_p^*, but denominator = (vol. of $(\mathbf{x}'\mathbf{V}_T^{-1}\mathbf{x} < c^2)$)
Chan *et al.* (1990)	C_{pm}	As C_p^* but denominator $E[(\mathbf{X} - \mathbf{T})'\mathbf{V}_0^{-1}(\mathbf{X} - T)]$
Pearn *et al.* (1992)	$_\nu C_{pm}$	$_\nu C_p \times C_{pm}$
Proposal 1.	$_\nu C_R$	$\|\mathbf{V}_0 \mathbf{A}^{-1}\| + (\xi - \mathbf{T})'\mathbf{A}^{-1}(\xi - \mathbf{T})$
Proposal 2.	$E[L(\mathbf{X})]$	$\text{tr}\,(\mathbf{A}^{-1}\mathbf{V}_0) + (\xi - \mathbf{T})'\mathbf{A}^{-1}(\xi - \mathbf{T})$

$\mathbf{V}_0$ is the variance-covariance matrix of the joint distribution of the process characteristics, and $\mathbf{V}_T = \mathbf{V}_0 + (\xi - \mathbf{T})(\xi - \mathbf{T})'$.

where $h(\cdot)$ is a nonnegative scalar function, satisfying the condition $h(t\mathbf{x}) = th(\mathbf{x})$ for all $t > 0$. Then a multivariate PCI can be defined as

$$MC_p = r/r_0$$

where

$$\Pr[h(\mathbf{X} - \mathbf{T}) > r] = \alpha$$

This ensures that if $MC_p = 1$ the expected proportion of NC items is α.

This definition includes, for example, the rectangular specification region

$$|X_j - T_j| < r_j \quad (j = 1, \ldots, \nu)$$

by taking

$$h(\mathbf{X} - \mathbf{T}) = \max_{1 \leqslant j \leqslant v} r_j^{-1} |X_j - T_j|$$

and $r_0 = 1$.

In this case, and several others, actual estimation of MC_p (i.e. of r) can entail quite heavy computation. Some details of possible methods of calculation are provided by Chen (1992).

4. Wierda's (1992a, b) suggestion that $-\frac{1}{3} \Phi^{-1}$ (expected proportion (p) of NC items) be used as a PCI, is applicable in multivariate, as well as univariate situations. It can then be regarded as a multivariate generalization of C_{pk}. The same comments – that one might as well use p, itself, and that p is estimated on the basis of an assumed form of process distribution, accuracy will depend on correctness of this assumption – apply, as mentioned in section 2.

5.5 VECTOR-VALUED PCIs

As we have noted (several times) it is a risky undertaking to represent variation of even a univariate characteristic by a single index. The possibility of hiding important information is much greater when multivariate characteristics are under consideration, and the desirability of using vector valued PCIs arises quite naturally. One such vector would consist of the v PCIs, one for each of the v variables. These might be C_ps, C_{pk}s or C_{pm}s, and would probably be related to the concept of rectangular parallelopiped specification regions.

We have already mentioned (in section 5.1) a suggestion to use the product of C_{pm}s, and have pointed out the deficiencies of this proposal. Interpretation of the set of estimated PCIs would be assisted by knowledge of the joint distribution. The work of Kocherlakota and Kocherlakota (1991) provides the joint distribution of C_ps for $v = 2$, under assumed multi-normality.

A different type of vector PCI has been proposed by Hubele *et al.* (1991), in which they suggest using a vector with three components.

1. The ratio

$$\left[\frac{\text{area of specification rectangular paralellopiped}}{\substack{\text{area of smallest rectangle containing the } 99.73\% \\ \text{ellipsoid, } (\mathbf{x} - \boldsymbol{\xi})' \mathbf{V}_0^{-1} (\mathbf{x} - \boldsymbol{\xi}) \leqslant \chi_{v, 0.9973}^2}} \right]^{\frac{1}{v}}$$

$$= \left[\frac{\text{'specification rectangle'}}{\text{'process rectangle'}} \right]^{\frac{1}{v}} \tag{5.29 a}$$

The numerator is (see (5.10))

$$\left[\prod_{i=1}^{v} (\text{USL}_i - \text{LSL}_i) \right]^{\frac{1}{v}}$$

Hubele *et al.* (1991) show that the denominator is

$$2\chi_{v, 0.9973} \left(\left\{ \prod_{i=1}^{v} |\mathbf{V}_{0ii}^{-1}| \right\}^{\frac{1}{v}} \Big/ |\mathbf{V}_0^{-1}| \right)^{\frac{1}{2}} \tag{5.29 b}$$

where, $\mathbf{V}_{0ii}^{-1}$ is obtained from $\mathbf{V}_0^{-1}$ by deleting the *i*th row and the *i*th column.

2. The *P*-value of the Hotelling T^2 statistic

$$W^2 = n(\bar{\mathbf{X}} - \mathbf{T})' \hat{\mathbf{V}}^{-1} (\bar{\mathbf{X}} - \mathbf{T}) \tag{5.30}$$

where $\hat{\mathbf{V}}$ is the sample variance–covariance matrix. This component includes information about the *relative* location of the process and specification values. If $\boldsymbol{\xi} = \mathbf{T}$ – i.e.

the process is accurately centred – then W^2 has the distribution of

$$\frac{v(n-1)}{n-v}\{F_{v,n-v} \text{ variable}\}$$

When $\bar{X}$ is close to T the P-value of W^2 is nearly 1; the further $\bar{X}$ is from T, the nearer is the P-value to zero.

3. The third component measures location and length of sides of the 'process rectangle' relative to those of the 'specification rectangle'. For the case $v = 2$, and using the symbols UPR_i, LPR_i to denote upper and lower limits for variable X_i in the process rectangle, the value of the component is

$$\max\left[1, \frac{|UPR_1 - LSL_1|}{USL_1 - LSL_1}, \frac{|LPR_1 - USL_1|}{USL_1 - LSL_1}, \frac{|UPR_2 - LSL_2|}{USL_2 - LSL_2}, \right.$$

$$\left. \frac{|LPR_2 - USL_2|}{USL_2 - LSL_2} \right]$$

A value greater than 1 indicates that part or whole, of the process rectangle falls outside the specification rectangle.

Taam *et al.* (1991) point out that components (1.) and (2.) of this triplet are similar to the numerator (C_p) and the estimator of the denominator,

$$[1 + (\bar{X} - T)'\hat{V}^{-1}(\bar{X} - T)]^{\frac{1}{2}} = [1 + n^{-1}W^2]^{\frac{1}{2}}$$

of C^*_{pm} (see (5.18)), respectively.

Although these three components do measure different aspects of process capability, they are by no means exhaustive. Also, when $v > 2$, calculation of component (3.) is cumbersome, and more importantly, there is no practically relevant distribution theory. Nevertheless, (1.), (2.) and (3.), combined will give a useful idea of the process capability – especially indicating in what respects it is, and is not, likely to be satisfactory.

APPENDIX 5.A

We investigate the distribution of

$$W = (\mathbf{X} - \mathbf{T})'\mathbf{A}^{-1}(\mathbf{X} - \mathbf{T})$$

where $\mathbf{X}$ has a v-dimensional multinormal distribution with expected value vector $\boldsymbol{\xi}$ and variance–covariance matrix $\mathbf{V}_0$, and $\mathbf{A}$ is a positive definite $v \times v$ matrix.

From Anderson (1984, p. 589, Theorem A2.2) we find that there exists a nonsingular $v \times v$ matrix $\mathbf{F}$ such that

$$\mathbf{F}\mathbf{V}_0\mathbf{F}' = \mathbf{I} \qquad (5.31\,a)$$

and

$$\mathbf{F}\mathbf{A}\mathbf{F}' = \operatorname{diag}(\lambda) \qquad (5.31\,b)$$

where $\operatorname{diag}(\lambda)$ is a $v \times v$ diagonal matrix in which the diagonal elements $\lambda_1, \ldots, \lambda_v$ are roots of the determinantal equation

$$|\mathbf{A} - \lambda\mathbf{V}_0| = 0 \qquad (5.32)$$

and $\mathbf{I} = \operatorname{diag}(\mathbf{1})$ is a $v \times v$ identity matrix.

Note that (5.31a) can be rewritten

$$\mathbf{F}'\mathbf{F} = \mathbf{V}_0^{-1} \qquad (5.31\,c)$$

and (5.31b) as

$$(\mathbf{F}^{-1})'\mathbf{A}^{-1}\mathbf{F}^{-1} = \operatorname{diag}(1/\lambda) \qquad (5.31\,d)$$

If $Y = F(X - \xi)$ then Y has a v-dimensional multinormal distribution with expected value O and variance–covariance matrix

$$E[YY'] = FE[(X - \xi)(X - \xi)']F' = FV_0F' = I$$

from (5.31a). Then since $X - \xi = F^{-1}Y$,

$$
\begin{aligned}
W &= (F^{-1}Y + \xi - T)'A^{-1}(F^{-1}Y + \xi - T) \\
&= (Y' + (\xi - T)'F')(F^{-1})'A^{-1}F^{-1}(Y + F(\xi - T)) \\
&= (Y + F(\xi - T))' \operatorname{diag}(1/\lambda)(Y + F(\xi - T)) \\
&= \sum_{j=1}^{v} \lambda_j^{-1}(Y_j + \delta_j)^2
\end{aligned}
\tag{5.33}
$$

where $\delta' = (\delta_1, \ldots, \delta_v) = (\xi - T)'F'$.

We note that in the particular case when $A = \theta^2 V_0$ we have from (5.32),

$$\lambda_1 = \lambda_2 = \cdots = \lambda_v = \theta^2$$

so

$$W = \theta^{-2} \sum_{j=1}^{v} (Y_j + \delta_j)^2 \tag{5.34}$$

and is distributed as

$\theta^{-2} \times$ (noncentral chi-squared with v degrees of freedom and noncentrality parameter

$$\sum_{j=1}^{n} \delta_j^2 = \delta'\delta = (\xi - T)'F'F(\xi - T) = (\xi - T)'V_0^{-1}(\xi - T).$$

If we have $T = \xi$, i.e. the process is centred at the target value, then $\delta = 0$ and W reduces to

$$W = \sum_{j=1}^{n} \frac{Y_j^2}{\lambda_j} \tag{5.35}$$

(W is the weighted sum of n independent χ_1^2 variables.)

The general distribution of W in this case has been studied by Johnson and Kotz (1968) *inter alia*. They provide tables of percentage points of the distribution for $v = 4$ and $v = 5$ for various combinations of values of the λs subject to the condition

$$\sum_{j=1}^{v} \frac{1}{\lambda_j} = v$$

These $100(1-\alpha)\%$ percentage points are values of c_v^2 such that $\Pr[W < c_v^2] = 1 - \alpha$. From the tables one can see how c_v^2 varies with variation in the values of the λs.

Some values for $v = 4$ are shown in Table 5.2.

Table 5.2 Percentage points of W

λ_1^{-1}	λ_2^{-1}	λ_3^{-1}	λ_4^{-1}	97.5%	99%	99.5%	99.73%
2.5	0.7	0.4	0.4	14.3	18.3	21.4	24.1
2.0	1.0	0.7	0.3	12.8	16.0	18.4	20.7
1.5	1.5	0.8	0.2	12.4	15.2	17.3	19.2
1.5	1.5	0.5	0.5	12.3	15.0	17.1	19.0
1.2	1.2	1.2	0.4	11.7	14.1	15.9	17.5
*1.0	1.0	1.0	1.0	11.1	13.3	14.9	16.25

In all cases $\Sigma_{j=1}^{4} \lambda_j^{-1} = 4$.
*The values in this row are percentage points of χ_4^2 – chisquared with four degrees of freedom.

The more discrepant the values of the λs, the greater the correct value of c_v^2. This will, of course, decrease the values of the PCIs defined in sections 5.3 and 5.4, implying a greater proportion of NC items than would be the case if all λs were equal to 1. Of course, one must keep in mind that this conclusion is valid with regard to situations restricted by the requirement of a fixed value of the sum

$$\sum_{j=1}^{n} \frac{1}{\lambda_j}$$

If all the λ_js are larger in the same proportion, one would obtain proportionately smaller values of c_v^2.

BIBLIOGRAPHY

Alt, F.B. (1985) Multivariate quality control. In *Encyclopedia of Statistical Sciences*, (S. Kotz, N.L. Johnson and C.B. Read, eds.) **6**, Wiley: New York, 110–22.

Anderson, T.W. (1984) *An Introduction to Multivariate Statistical Analysis* (2nd edn) Wiley, New York.

Chan, L.K. (1992), Cheng, S.W. and Spiring, F.A. (1990) A multivariate measure of process capability, *J. Modeling Simul.*, **11**, 1–6. [Abridged version in *Advances in Reliability and Quality Control* (1988) (M. Hamza, ed.) ACTA Press, Anaheim, CA; 195–9].

Chen, H.F. A multivariate process index over a rectangular paralellopiped, *Tech. Rep.* Math. Statist. Dept, Bowling Green State University, Bowling Green, Ohio.

Hotelling, H. (1947) in *Techniques of Statistical Analysis* (C. Eisenhart, H. Hastay and W.A. Wallis, eds.) McGraw-Hill: New York, 111–84.

Hsiang, T.C. and Taguchi, G. (1985) *A Tutorial on Quality Control and Assurance – The Taguchi Methods*, Amer. Statist. Assoc. Meeting Las Vegas, Nevada (188pp.).

Hubele, N.F., Shahriari, H. and Cheng, C.-S. (1991) A bivariate process capability vector, in *Statistical Process Control in Manufacturing* (J.B. Keats and D.C. Montgomery, eds.) M. Dekker: New York, 299–310.

Johnson, N.L. and Kotz, S. (1968) Tables of the distribution of quadratic forms in central normal variables, *Sankhyā Ser. B*, **30**, 303–314.

Johnson, T. (1992) The relationship of C_{pm} to squared error loss, *J. Qual. Technol.*, **24**, 211–15.

Kocherlakota, S. and Kocherlakota, K. (1991) Process capability indices: Bivariate normal distribution, *Commun. Statist. – Theor. Meth.*, **20**, 2529–47.

Littig, J.L., Lam, C.T. and Pollock, S.M. (1992) Process capability measurements for a bivariate characteristic over an elliptical tolerance zone, *Tech. Rep.* 92–42, Dept. Industrial and Operations Engineering, University of Michigan, Ann Arbor.

Pearn, W.L., Kotz, S. and Johnson, N.L. (1992) Distributional and inferential properties of process capability indices, old and new, *J. Qual. Technol.* **24**, 215–31.

Taam, W., Subbaiah, P. and Liddy, J.W. (1991) A note on multivariate capability indices. Working paper, Dept. Mathematical Sciences, Oakland University, Rochester MI.

Wierda, S.J. (1992a) *Multivariate quality control: estimation of the percentage good products.* Research memo, No. 482, Institute of Economic Research, Faculty of Economics, University of Groningen, Netherlands.

Wierda, S.J. (1992b) A multivariate process capability index, *Proc. 9th Internat. Conf. Israel. Soc. Qual. Assur.* Jerusalem, Israel.

Note on computer programs

The following SAS programs, written by Dr Robert N. Rodriguez of SAS Institute Inc., SAS Campus Drive, Cary NC 27513–8000, USA, will be available in the SAS Sample Library provided with SAS/QC software (Release 6.08, expected in early 1993). A directory of these programs is given in the Sample Library program named CAPDIST.

Chapter 2
Expected value and standard deviation of $\hat{C}_p$
Surface plot for bias of $\hat{C}_p$
Surface plot for mean square error of $\hat{C}_p$
Expected value and standard deviation of $\hat{C}_{pk}$
Surface plot for bias of $\hat{C}_{pk}$
Surface plot for mean square error of $\hat{C}_{pk}$
Plot of probability density function of $\hat{C}_p$
Plot of probability density function of $\hat{C}_{pk}$
Confidence intervals for C_p
Approximate lower confidence bound for C_{pk} using the approach of Chou *et al.* (1990) (reproduces their Table 5)
Approximate lower confidence bound for C_{pk} using the approach of Chou *et al.* (1990) and the numerical method of Guirguis and Rodriguez (1992) (generalizes Table 5 of Chou *et al.*)
Approximate confidence intervals for C_{pk} using the methods of Bissell (1990) and Zhang *et al.* (1990)

Chapter 3
Expected value and standard deviation of $\hat{C}_{pm}$
Expected value and standard deviation of $\hat{C}_{pmk}$
Surface plot for bias of $\hat{C}_{pm}$
Surface plot for mean square errr of $\hat{C}_{pm}$
Approximate confidence limits for C_{pm} using the method of Boyles (1991)

Chapter 4
Expected value and standard deviation of $\hat{C}_{jkp}/C_{jkp}$
Surface plot for bias of $\hat{C}_{jkp}$
Surface plot for mean square error of $\hat{C}_{jkp}$

Postscript

Our journey down the bumpy road of process capability indices has come to an end, for the moment. We hope that those who have patiently studied this book, or even those who have browsed through it in less detail, have been motivated to seek for answers to the following questions.

1. Are the resistance and objections to the use of PCIs in practice justified?
2. Is there truth in the statement that 'a manager would rather use a wrong tool which (s)he understands than a correct one which (s)he does not'?
3. Are PCIs too advanced concepts for average technical workers on the shop floor to comprehend?

Let us state immediately that our answer to the last question should be an emphatic 'No'. In our opinion it is definitely not too much to expect from workers a basic understanding of elementary statistical and probabilistic concepts such as mean and standard deviation (and target value and variability in general).

We also sincerely hope that the answer to the second question is an equally emphatic 'No'. In spite of rather disturbing catch phrases and slogans, such as 'KISS (keep it simple, statisticians)' and even 'Statisticians keep out' (ascribed to Taguchi advocates at a meeting in London), we strongly believe that a great majority of managers and engineers are sufficiently enlightened and motivated to explore new avenues, and are not lazy, or afraid of the 'unknown'. Deming's philosophy and a shrinking world (from development of 'instant' communications) are factors contributing to an atmosphere favourable to development of innovative techniques on the shop floor, and improvements in educational methods may be expected to produce many more open-minded individuals.

Having explained our reactions – we hope, satisfactorily – to the second and third questions we now attempt to reply to the first, and most difficult, one. We rephrase the question – should the use of PCIs be encouraged?

As we indicated in the introduction to Chapter 1, PCIs can be used with advantage provided we understand their limitations – in particular, that a PCI is only a single value and should not usually be expected to provide an adequate measure on its own, and PCIs are subject to sampling variation, as are other statistics. In the case of sample mean $(\bar{X})$, for example, the sampling distribution of $\bar{X}$ is usually approximately normal even for moderate sample sizes ($n > 6$, say), even if the process distribution is not normal. We therefore use a normal distribution for $\bar{X}$ with some confidence. On the other hand none of the sampling distributions of PCIs in this book is normal, even when the process distribution is perfectly normal.

Tables of the kind included in this book provide adequate background for interpreting estimates of PCIs when the process distribution is normal. In our opinion, objections, voiced by some opponents of PCIs, that are based on noting that the distribution of estimates of PCIs when the process distribution is non-normal is unknown and may hide unpleasant surprises are somewhat exaggerated. The results of studies, like those of Price and Price (1992) and Franklin and Wasserman (1992) summarized in Chapter 4 provide some reassurance, at the same time indicating where real danger may lurk. In particular the effects of kurtosis ($\beta_2 < 3$, or more importantly $\beta_2 \gg 3$) are notable.

We appreciate greatly the efforts of those involved in these enquiries, although we have to remark – with all due respect – that their methods (of simulation and bootstrapping) do not, as yet, provide a sufficiently clearcut picture for theoreticians to guess the exact form of the sampling distribution, on the lines of W.S. Gosset ('Student')'s classical work on the distribution of $\bar{X}/S$ in 1908 wherein hand calculation coupled with perceptive intuition led to a correct conjecture, deriving the t-distribution.

Index

212   *Index*